TECHNOLOGY MANAGEMENT AND PUBLIC POLICY IN THE EUROPEAN UNION

EUROPEAN COMMISSION

Technology Management and Public Policy in the European Union

Edited by

WILLIAM CANNELL and BEN DANKBAAR

OFFICE FOR OFFICIAL PUBLICATIONS OF THE EUROPEAN COMMUNITIES
and
OXFORD UNIVERSITY PRESS
1996

This book has been printed digitally and produced in a standard design in order to ensure its continuing availability

OXFORD
UNIVERSITY PRESS

Great Clarendon Street, Oxford OX2 6DP

Oxford University Press is a department of the University of Oxford.
It furthers the University's objective of excellence in research, scholarship,
and education by publishing worldwide in

Oxford New York

Athens Auckland Bangkok Bogotá Buenos Aires Cape Town
Chennai Dar es Salaam Delhi Florence Hong Kong Istanbul Karachi
Kolkata Kuala Lumpur Madrid Melbourne Mexico City Mumbai Nairobi
Paris São Paulo Shanghai Singapore Taipei Tokyo Toronto Warsaw

with associated companies in Berlin Ibadan

Oxford is a registered trade mark of Oxford University Press
in the UK and in certain other countries

Published in the United States
by Oxford University Press Inc., New York

Reprinted 2001

ISBN 0-19-829028-4

Preface

It has become increasingly clear that the competitive position of firms in advanced economies—indeed of economies in their entirety—depends not just on research activities and the technological capabilities they have at their disposal but on their capacity to manage technology. Although the literature on technology management is expanding, up to now very little has been published which explicitly considers the public policy implications of technology management and the capabilities and deficits of firms. The aim of this book is to go some way towards filling this gap and to address systematically the policy issues surrounding technology management in firms, so as to bring them to the attention of researchers and policy-makers.

The book is based on research carried out during a major empirical study conceived and supervised by the European Commission (DGXII—Science, Research and Development) within the framework of the MONITOR-SAST Activity (Strategic Analysis in Science and Technology), and carried out over a two-year period by a consortium of researchers from across the EU under contract no. MOSA-CT92-0045. It provides a conceptual foundation for linking the realities of technology management in European firms with current industrial and technology policy preoccupations, and presents an empirical analysis of technology management problems and associated policy needs, with respect to different types of firms and to the research and technology infrastructure in Europe.

The focus of the book is on *the firm-level competences underlying technology management*, and the policy instruments which can be used to promote them. The various chapters reflect the outcome of a systematic exploration of the diversity of technology management problems faced, and approaches adopted, within European companies. At the outset it has been assumed that the competitive performance of firms is strongly correlated with their technology management capabilities. But this is not a book about industrial competitiveness, which is conditioned by many other factors affecting the broader economic, institutional and social context of innovation outside the firm, and which are beyond the scope of our analysis.

The book is aimed at researchers and practitioners concerned with innovation and public policy in a European context. It has particular relevance for researchers with an interest in the comparative analysis of management practice and the contribution of management to industrial and technological change; and for policy analysts concerned with management as a factor in the development of industrial innovation, research and regional policies. It will also be of interest to a wider circle of researchers and professional groups concerned with industrial development in the context of European integration.

Obviously, this book could not have been written without the help of many people besides the authors. We have relied extensively on the interviews and case studies carried out by other members of the research consortium and have benefited greatly from their insights as participants in workshops conducted during the course of the study. We have also been greatly assisted by the contributions of a specially constituted panel of industrial experts who provided reviews of technology management problems and issues in various industrial and service sectors, and commented extensively on reports produced by the consortium. A full list of members of the research consortium and panel of experts, and a list of all the reports published from the study appear in appendices at the end of the book. The authors would also like to express their gratitude to the numerous representatives of enterprises who graciously gave of their time to participate in interviews and answer the numerous questions we had lined up for them in our common questionnaire.

The study was overseen by a Steering Group consisting of representatives of the services of the European Commission, who reviewed draft reports and contributed many helpful suggestions. Special thanks are due to Bruno Schmitz, head of the Monitor-SAST Activity, for his unfailing support for the study and his astute intellectual input. First drafts of the chapters of this book were presented and discussed at a workshop held in Brussels in November 1994, and have been much improved as a result. We would particularly like to thank the discussants at that workshop, whose names also appear in an appendix. Finally, this book would never have reached the publishers without the patience and diligence of our secretaries, Claire McCabe, Silvana de Sanctis, and during an earlier period Ellis Campbell, and we thank them for their cheerful support.

W.P.C.
B.D.

Contents

List of Figures

List of Tables

Notes on Contributors

WILLIAM CANNELL works on Technology Policy at the European Commission, DGXII (Science, Research, and Development) where he has been responsible for several major research studies on technological innovation. He has a degree in natural sciences and a Ph.D. in applied psychology. Previously he was Branch Manager in the Operational Research Directorate at the Civil Aviation Authority in London, dealing amongst other things with the appraisal of technology and organizational systems for managing safety in air traffic control.

ROD COOMBS is Professor of Technology Management at the Manchester School of Management, UMIST. His research and teaching activities are in the areas of the process of technological innovation at firm level; the economic and social analysis of technological change; and the role of technology in strategic management of enterprises and of countries. He is co-Director of the newly created ESRC Manchester Research Centre on Innovation and Competitiveness. He consults widely on technology-related issues for companies and government agencies.

BEN DANKBAAR is Professor of Business Administration at the Business School of the University of Nijmegen. He studied social sciences and economics at the University of Amsterdam and received a Ph.D. from the University of Limburg. He was co-ordinator of the international consortium for the MONITOR-SAST Study, which formed the basis for this book. At that time, he was programme leader at the Maastricht Economic Research Institute on Innovation and Technology (MERIT).

STEFAN KUHLMANN studied political science and history at the University of Marburg and gained his Ph.D. at the University—Gesamthochschule Kassel (Dr.rer.pol), where he carried out research in the field of the informatization of the public administration and the service sector. He joined the Fraunhofer Institute for Systems and Innovation Research (ISI) in 1988 and is now head of the Technology Assessment and Innovation Strategies Department. His present work is in the analysis, planning, and evaluation of research, technology, and innovation policies (national, international, regional), analysis of industrial innovation strategies, and technology assessment.

GUIDO REGER studied social and economic science at the Hochschule für Wirtschaft und Politik in Hamburg. From 1990 to 1991 he worked as a project manager at the Centre for Labour, Technics and Environment (ZATU) in Nuremberg. Since 1991 he has been working as project manager at the Fraunhofer Institute for Systems and Innovation Research (ISI), Karlsruhe, in the Technology Assessment and Innovation Strategies Department. His main fields of interest are industrial innovation strategies, technology management of enterprises, and the evaluation of international, national and regional technology policies.

LENA TSIPOURI is a professor at the University of Athens, in the Department of Economic Studies, where she teaches in the field of economic development and technological change in particular. She has studied economics and management in the Universities of Athens and Vienna, obtained her Doctorat d'État in Paris and has undertaken post-doctoral research at MIT under a Fulbright grant. Her academic publications and consulting work are concentrated in the areas of technology transfer and technology management.

1

Introduction: *Technology Management as a Focus for Public Policy*

William Cannell

Why does Technology Management Matter to Policy?

Technology management, in our definition, is the set of activities undertaken by firms to create, access, and mobilize technology in support of their business aims. It has become increasing relevant to policy-makers as they have come to recognize the complexity of innovation processes and the central role of the firm in driving technological change.

In the past several decades important changes have taken place in the landscape of science, technology, and industry in many countries of the world, including those of the European Union. Policy-makers and managers alike have been forced to develop new ways of thinking and acting to make the best of opportunities offered by technological change. New 'articles of faith' have been developed to underlie technology policies which, along with the institutions that support them, have evolved significantly in these countries over the last two decades. These articles of faith are strikingly similar to those which underlie the emergence of technology management as a business concept. Common to both is the notion of technological change as a complex, decentralized, and non-linear process, a process that is fundamentally linked to competitive behaviour and which is manifested mainly through the profit-seeking activities of firms, and a process in which the role of technology must be considered alongside many other factors, organizational, institutional, social, and economic.

The views expressed in the chapter are those of the author and should not be taken to represent the position of the European Commission.

For the individual firm, technology management provides the link between, on the one hand, the scientific and technological resources—either available within the firm or accessible on external markets—and, on the other, its long-term competitiveness. Since national competitiveness is a direct function of the competitiveness of the nation's firms, technology management is, in aggregate, a significant part of the link between the scientific and technological resources available to the economy—local, regional, national, or international—and that economy's competitiveness.

Herein lies the first reason for analysing technology management in firms from the perspective of public policy: to identify the key resources on which firms depend for their technological competitiveness, and to direct policies towards the areas in which deficits are found. One of the most basic and widely accepted functions of technology policies is to rectify market failures in the economy of science and technology. But market failures, we know, are ubiquitous: policies reflect a set of priorities chosen by the public authorities with regard to the need for different types of resources: skilled personnel, codified scientific results, training opportunities, technology-based equipment and materials, innovation finance, etc. They demand an understanding of where the optimal balance of resources lies for the population of firms they address. This is an empirical question, the answer to which depends upon the types of firms and their approach to technology management.

A second and closely connected reason for studying technology management from a policy perspective is to understand what *capabilities* are involved at the level of firms; capabilities which in aggregate clearly represent valuable economic assets for regions and nations. Such capabilities themselves depend on the more elementary technological resources available to firms, and in turn constitute resources on which a technology management strategy (implicit or explicit) can be built. Studying technology management empirically can help us to understand the nature of these capabilities, the extent to which they are specific to individual firms or generic across groups of similar firms, how they are built up and deployed by firms, and on what features of the external environment they depend. This can provide the basis for policies designed to improve technology management capabilities and thus leverage more effectively the available technological resources to provide increased social and economic gains.

A third reason for studying technology management in firms arises from the observation that, whilst technology policies can be interpreted as policies to manage technology *for the benefit of society as a whole*, in reality, society's management of technology is to a large extent, and inevitably, delegated to firms. The functions of both creating technologies and embedding them into a societal context are largely effected by firms: through long-term research which opens technological opportunities; through product development, commercialization and marketing, affecting what government, business and consumers can buy; through industrial process development and upgrading, which determine conditions of work; through incorporation of technologies into business processes, which affects the types of services available and the ways in which they are delivered. Policies do not directly influence technological change, they only do so indirectly by influencing, amongst other things, the behaviour of firms. Unless we understand the ways in which the technological decisions of firms are made, and the constraints which bear upon these decisions, it is very difficult to understand the ways in which policies can most effectively or reliably influence them.

Finally, and without wishing to overstretch the analogy between the public and private spheres, another benefit to policy of studying technology management arises from the possible application of technology management approaches and methods in the policy domain. Continuing budgetary pressures are forcing national authorities in the EU and elsewhere to examine very carefully the disciplinary and sectoral focus of their support to technology, to set research priorities in accordance with the most favourable opportunities for national competitive advantage, and to develop an optimal set of interactions between the research base and industrial 'users'. These are not a long way from the problems faced by large multi-technology firms in defining their preferred areas of technological specialization, orienting their future research efforts, and making the most effective use of their current resources. It is at least plausible that the methods and approaches adopted in firms to address these problems could be strongly relevant to the policy domain.[1]

[1] As will become clear in what follows, although this study sheds some light on all of these domains, we do not pretend to have addressed all of them in a comprehensive manner.

Technology Policy and its Links with Technology Management

Technology Policy in Transition

Most advanced countries have established sets of policies which are directed towards the national capacity to produce and utilize technology in the interests of social and economic development. These include, most notably, incentives such as subsidies and/or tax concessions for firms investing in R&D or hiring research and engineering personnel (OECD, 1992*a*). Although their roots can clearly be traced to science policies and they still share some of the same types of instruments, technology policies are now considered to fall into a specific and separate policy area. They tend to embody distinctive objectives (most notably the advancement of industrial competitiveness) and address particular problems (such as the uptake and utilization of technology by industrial firms). They reflect a perception that the world of technology, which is mainly the province of firms searching for profits, follows rules which are rather different from that of science, despite the fact that the two worlds are intimately associated and mutually dependent on one another.

An analysis of developments in technology policies faces a number of obstacles. First of all, their limits are difficult to define. A wide range of policies—it could indeed be argued most policies—have significant impacts on technology development, for example:

- education and training policies, which are strongly linked to technology policies through their influence on technology-related skills, and the fact that research is an important medium for training scientists and engineers;
- policies affecting the investment climate, trading conditions, and intellectual property, which have an influence on the incentives for innovation and access to technology produced abroad;
- regulatory policies, such as those in the fields of environment and health, which constrain the technology choices of regulated firms and the future development prospects of particular technologies.

As a result, policies influencing technology development have different objectives, which may sometimes be in conflict, and which are likely to change in complex ways as economic and social conditions evolve. Moreover, the institutional and cultural differences between countries, the differences in their industrial heritage and development history, their national preoccupations and the ideological preferences of their governments, all contribute to considerable variation in the particular sets of policies implemented in different countries at any given time. A further difficulty derives from the many uncertainties underlying technology policies, including a serious lack of relevant data about the ways in which firms are responding to the challenges of technology and inadequate theoretical models of innovation (Smith, 1994).

The theoretical discourse surrounding policy (which is characterized by different theoretical perspectives and points of view about the determinants and consequences of technological change and innovation) is connected, but not in a simple way, with the development of policies themselves (which reveal large differences resulting from historical and political factors). Without going into a detailed discussion of the evolution of technology policies (see e.g. Rothwell and Dodgson, 1992; Ruivo, 1994; Schienstock, 1994; all of whom have sketched out various phases of technology policy during the past several decades), it is nevertheless possible to identify some important aspects of policy thinking in recent years, which help to explain some of the characteristic features of present policies and their relevance to technology management.

Current approaches to policy have been shaped by radical changes over the last few decades in both the pace of technological change and the impact this has had on industry and society. Fuelled by highly visible developments in electronic and information technology, there is now a widespread perception that a technological revolution is taking place, or indeed a series of overlapping revolutions involving information and communications technologies, materials technologies, and technologies arising from the life sciences. An earlier period of relatively steady and predictable technological advance has been brought to an end, and replaced with much more pervasive and interconnected changes, which have highly disruptive effects as they are accommodated by industry and society. The new technology has a different paradigmatic character, and its successful development and implementation involves new economic approaches.

This interpretation has figured strongly in attempts to explain the major macroeconomic problems which have troubled the western economies since the 1960s, notably the decline in productivity and the surge in levels of unemployment (OECD, 1994). A major feature of modern technology policies—as indeed of industrial policies more generally—is their explicit orientation not towards technology *per se* but towards managing the structural transitions at all levels in the economy and society which are necessary to accommodate, and to facilitate, technological change. Behind such an approach is the notion that a 'comprehensive innovation process' is needed, to accommodate to new rules of the game in which 'entrepreneurship, management, work organization, financing, the opening up of markets; worker and employer organizations, labour markets, educational and regional authorities, etc. are all involved' (OECD, 1988*a*: 11).

Contributing greatly to this interpretation has been the replacement of neo-classical approaches with evolutionary approaches as the dominant tendency in economic theories of technological change (Metcalfe, 1995). The firm emerges as the central actor from such a perspective. Evolutionary theories see technology not as an external factor in the economy, available on equal terms to all economic actors, but as the outcome of highly decentralized competitive behaviour on the part of firms, each of which has a different set of capabilities and opportunities. On this view, if technology is to contribute to social welfare, it does so not because of autonomous research activities supported directly by the state but because the profit-seeking activities of firms lead them to produce new or improved technology-based products and services, some of which are successful enough to supplant earlier versions. The evolutionary perspective also explains why policy-makers should be concerned with, on one hand, ensuring that the variety of innovation offerings is maximized and, on the other, that the 'framing conditions' are in place to ensure that market mechanisms select technologies which are beneficial to long-run as well as short-run social goals (e.g. environmental protection). This is one reason for the focus of modern technology policies on improving industrial competitiveness, the presumption being that this is closely equivalent to optimizing firms' abilities to develop and deploy technologies in the interests of successful innovation.

Within this framework, policy has come to depend on an increasingly sophisticated understanding of the factors—both internal and external—which influence the behaviour of firms with respect to technology, and their corresponding competitive ability. It is no longer possible for policy-makers to see the firm in classical economic guise as an actor responding 'rationally' to economic signals, with perfect information and an infinite set of options at its disposal to maximize profit in any decision situation. Instead, it must be acknowledged that firms' outlook, strategy, and day-to-day activities are subject to cognitive and organizational constraints (March and Simon, 1993) as well as those arising from their existing technological capabilities (Pavitt, 1992) and other assets (Teece, 1988). Because technological capabilities depend upon complementary and mutually reinforcing investments in knowledge and skill, much of it tacit, they accumulate only gradually, in a firm-specific way, and strongly influence firms' perception of future possibilities, as well as their capacity to take advantage of them. The quickening pace of technological change, combined with larger markets and stronger competition, also change radically the criteria for effective firm organization. Learning and adaptation to change, rather than 'static' efficiency, are the new desiderata, favouring flatter organizational structures, co-operative, decentralized management and a multi-functional, multi-skilled workforce (OECD, 1992*b*; Garvin, 1993).

Regarding the external factors, a major influence on modern technology policies is the recognition that technological innovation is crucially influenced by structural characteristics of the economy, and that innovation is itself fundamentally systemic in nature. There is a division of labour between the various types of organization which are involved in technological development—firms, universities, contract research organizations, professional organizations, sectoral research institutes, and so on—each of which contributes a different type of knowledge, which is accumulated through specific mechanisms and responds to specific types of incentives. Moreover, firms depend in various ways on other firms—customers, suppliers, and competitors—for technological and market opportunities (Pavitt, 1984; von Hippel, 1988). The availability of partners and the nature and quality of their interactions with them is highly dependent on the industrial structures in which they operate (Porter, 1990), and other institutions, such as industrial relations and organizational traditions (Kogut, 1993). The concept of 'national systems of innovation' (Lundvall, 1992; Nelson, 1993) is perhaps the most sophisticated expression of these ideas.

Technology Management: the Emergence of the Concept

The concept of technology management draws on the same roots as contemporary technology policies. Reflecting the principles of technology policy into the private sphere, technology management sees technology as a core strategic resource, with many systemic characteristics, which must therefore be considered from an integrated perspective. On the part of firms, technology management serves to ensure that the synergies between different technological capabilities and applications are fully exploited, and that the organization, skills, and other associated factors are in place to optimize the value of these resources. This obviously requires a sophisticated understanding of the nature of technology and the determinants of technological change.

Like business strategy more generally, technology management can be seen as a function at the interface between a series of bottom-up and top-down processes of organizational development, both of which serve to emphasize the growing importance of technology as a component of the way business is done. Technology management is the point at which long-term visions of research and short-term (continuous) improvements in products and processes meet (see Figure 1.1).

The search for improved performance of existing business processes has, on the one hand, brought together methods and technologies that were previously separate: total quality management, just-in-time logistics, electronic data interchange, CAD-CAM, management information systems, etc. These partial improvements are being increasingly integrated into more comprehensive approaches requiring a strategic orientation over the enterprise as a whole: lean production, computer-integrated business, business re-engineering.

On the other hand, there is growing appreciation of the need to link research and development more closely to the long-term needs of the enterprise. This has led to a number of management approaches which link the firm's overall business strategy with its technological capabilities. The concepts of 'core competence' and 'core technology' are illustrative of this trend. Firms are increasingly thinking of technology not just as an internal factor which forms a major element of their specific competitive strengths, but as a means to leverage external capabilities, via strategic alliances and joint ventures.

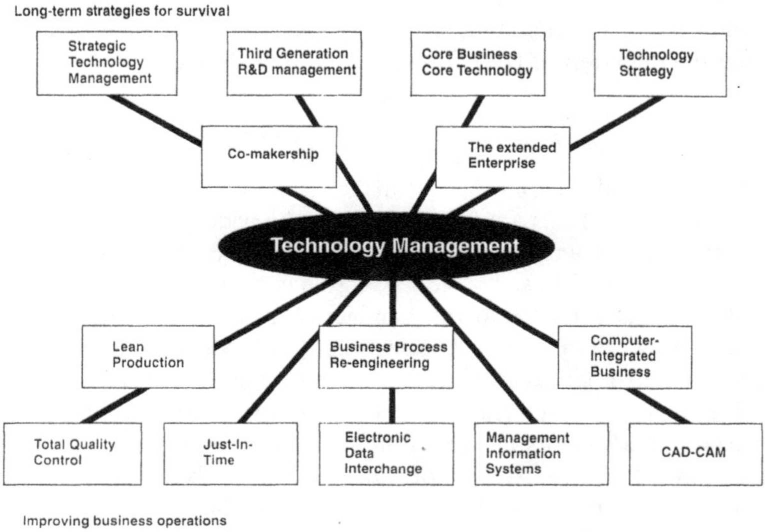

Fig. 1.1. Technology Management: where long-term visions and short-term improvements meet

An important impetus towards the technology-management approach comes from the lessons of failure in accommodating to technological change. Heavy investments in new production technology were made by firms in the 1980s with little understanding of their operational and strategic implications. They were driven by the perception that easy gains in productivity could be captured by introducing high technology automation such as robotics, and supported by various government initiatives. The widespread failure to realize the anticipated benefits has been well documented (Bessant, 1993), and the lessons learnt have been a determining factor in developing the more sophisticated approach to technology associated with the notion of technology management.

Technology management implies a movement from considering technology in functional terms or in relation to individual business processes, towards an overall appreciation of technology from a corporate perspective. It also reflects an appreciation that the fundamental and distinctive competences of firms are not static features which relate only to their present products and processes, but that they have important dynamic characteristics, at the same time allowing an accumulating innovative capability, but also constraining the direction of their future competitive development.

The Rationale for Technology Management

The essential factors which make technology management so important for firms are fourfold. First, the widespread and diversified impact of technology on competitive performance. In practically all areas of business, technology has become a dimension on which firms are able to differentiate themselves from their competitors, whether their strategies relate to price, quality, speed to market, or other aspects of competition. This is no less true of service than manufacturing sectors: the information revolution has ensured that sectors like telecommunications and financial services are amongst the most technologically dynamic. Neither is the phenomenon restricted to high-tech industries; the sectors just cited, for example, carry out relatively little research of their own, and there are innumerable examples of where astute introduction of new process technologies has rejuvenated industries once regarded as mature. Technology management fills the need for a much broader concept than research management, which is relevant to the complete range of processes within the firm and practically all types of firm.

Second, the interdependence between technology, knowledge, and skills. What the failures of the 1980s demonstrated above all was that the level of performance of complex technological systems is not an attribute of the hardware alone. Understanding and making use of the full potential of investments in technology depends upon complementary investments in the knowledge and skill of the people making use of the technology. This goes well beyond the workforce directly associated with the technology in question, and may affect all levels of management. The specific knowledge necessary to master complex technologies in a specific business context is rarely widely available, and accumulates only slowly. This means that firms are closely bound to the development paths that their previous history has prepared them for, and their business strategies need to be based on an understanding of the limitations of their own technological capabilities in the face of radical changes in the technological basis of their industries.

Third, the organizational implications of technological change. Exploiting in-house technological resources in the development of new products and processes depends upon an organizational capacity to transcend functional and divisional boundaries, and thus bring together research activities from different product areas and put them in close contact with development and marketing activities, etc. Exploiting external sources of technology, on the other hand, can demand a capability to transform production processes to take advantage of the new possibilities opened up by the technology. Emerging management paradigms such as 'lean enterprise' (Womack and Jones, 1994), 'core competences' (Prahalad and Hamel, 1990), 'business process re-engineering' (Johansson *et al.*, 1993), and so on, can be seen to involve both organizational and technological components, closely linked.

Fourth, the limits to technological autonomy enjoyed by firms. The complexity of modern technologies and their intrusion into all aspects of business mean that even the largest and most technologically advanced firms depend quite extensively on technology sourced from outside. All firms face the need to monitor technological developments undergone by their suppliers, customers, and competitors. Increasingly, firms are thinking of leverageing their own technological capabilities by combination with those of other firms, through strategic alliances, joint ventures, and more informal networking arrangements. This brings additional organizational and skill-related problems internally, to manage a variety of external interfaces and assimilate the knowledge arising from them, whilst protecting against leakage of the firm's own intellectual property.

Common Factors Underlying the Evolution of Policy and the Emergence of Technology Management

Technology management in firms and the new policy thinking on industrial competitiveness have emerged from similar concerns on the part of industry and government. They are both responses to a perceived failure to make full use of the potential offered by technological change. Both acknowledge the central role of the firm in technological change. And both reflect a realization that a new understanding of technology is needed, which is much more intricately linked to other processes of change occurring in firms and economies.

The picture of technological change which has developed out of these two developments is not that of a deterministic 'natural' process, external to, and independent of, other economic activities. Instead, it is heavily conditioned by the innovation processes which mediate the interface between technological possibility and market opportunity. Many layers of feedback and iteration are involved in moving a scientific breakthrough or technological idea towards industrial and commercial practice; more still in the diffusion of such practices within industry and society generally. The economic and social benefit to be derived from the technological capabilities of firms and nations therefore depends on the mobilization and organization of a wide range of other, non-technical, assets.

The main substantive link between policy and technology management arises from the fact that policy needs to be based on a behavioural understanding of the firm—the way in which management and organizational routines derive from, and simultaneously reinforce, technological capabilities and learning mechanisms. The accumulation of scientific and technological capabilities itself follows a logic which is closely connected to the dynamics of innovation within the firm. The ability of firms and nations to develop new technological capabilities depends crucially on those they already possess, embodied not just in research results and physical assets, but largely in the knowledge and skills of their workforce, and the organizational routines by which this intellectual capital is put to use in solving practical business problems. This is precisely what technology management is all about.

The Objectives of the Study and its Contributions to Policy

Although advances in understanding of innovation and technological change have led to a wholesale replacement of the articles of faith underlying technology policies, we do not yet have theories which provide clear guidance on how policies should be constructed. The reasons for this include the complexity of the processes involved, the diversity of situations on which policies need to act, and the fundamental indeterminacy of important elements of the process (i.e. scientific discoveries, technological opportunities, and market demands). Moreover, policy design must respond, not just to economic rationality, but to strictly political considerations, such as the balance between equity and efficiency, and to the administrative constraints associated with the use of particular instruments. In technology policies, as in other policies, there is evidence of strong continuity, which can be considered quite rational from the perspective of maintaining future options and avoiding the risk of unintended consequences.

In practice, therefore, technology policy formulation and implementation is rarely the direct consequence of theory, and relies on other types of information and incentives, as well as general principles. Edquist (1994) identifies imitation and lobbying as plausible drivers of policy change, and points out the importance of policy evaluation in the context of the specific characteristics and problems of the economic environment which policies are intended to 'fix'. This view holds that policy should evolve, probably on an incremental basis, in response to 'diagnostic analysis', which involves empirical and comparative, as well as explanatory and theoretical, elements. This is the spirit in which our study is conceived. We are adopting the concepts of modern innovation theory to characterize and explain, on a comparative basis, the technology management problems of European industrial firms and the ways in which they are being addressed. We are drawing conclusions regarding the key issues facing these firms and the weakest links in these processes. And we are drawing policy conclusions within the framework of Community RTD and other policies. The following types of question are indicative of the possible contributions of the study to contemporary policy problems.

What are the main resource bottlenecks affecting technology management?

The main objective of this study was to develop a better understanding of the technology management capabilities of firms within the EU and their dependences on the resources available to the firm either internally or accessed from outside. The idea was to analyse the site of the main bottlenecks in the resources and capabilities involved in technology management, for European firms, and the implications this has for technology policies. As we have seen, implicit in the concept of technology management is that a range of other resources and capabilities, beyond R&D, are involved in successfully developing and marketing technology-based products. If the bottlenecks for firms are not, in fact, limitations in research outputs, their competitive position is unlikely to be improved by policies intended to secure a greater supply of research alone. On the other hand, policies directed towards improving the availability of other, complementary, resources which are in short supply, may provide considerable leverage of existing research outputs.

How do the characteristics of firms determine their technology-management needs?

By systematically analysing the technology-management practices and problems of a large number of firms, distributed across a range of sectors and regions, the study allows us to identify some characteristics of firms which have the most important influence on their technology management needs. We emphatically reject the suggestion that there is one way to do technology management, for any group of firms; technology management has a strong strategic component, which differentiates the competitive approach of firms in the same industry. Equally, however, we recognize that policies need to address the areas of greatest need, whilst avoiding discrimination between firms. Designing policies therefore requires an understanding of how the various 'target groups' can be defined, and how their needs differ. With the prevailing philosophy that science and technology policies—like other economic policies—should be 'non-sectoral' so as to foster 'industrial change', it is especially important to understand what characterizes equivalent groups of firms from the perspective of technology management, on a cross-sectoral basis.

How do firms build up their technology management capabilities?

The study provides a number of indications regarding the ways in which technology-management competences are developed in firms: how firms learn about technologies, the constraints they face in doing so, and the ways in which these are affected by their strategic outlook and existing capabilities. These are important because, if policies directed towards improving technology-management resources are to have their desired effect on firms' capabilities, they must be designed with an understanding of how such capabilities are developed internally by firms or assimilated from outside. But policies devoted to improving the environment for innovation, or increasing access to technology have often paid little attention to the internal capabilities required by firms to make use of the services on offer.

Policies affect the environment of firms, by adjusting economic incentives and regulations, and changing the organization of technology and product markets. A key question is therefore how enterprises respond to environmental changes by their technology-management strategies and decisions (in the short term) and with regard to the development of competences (in the long-term). Another issue relates to the way in which firms learn from their own experience and transfer knowledge within the organization so as to make it efficiently available where needed. The intangible nature of technology-management capability means that these have to be understood as learning processes, not as simple transactions between the policy sphere and the firm. This raises many questions, not least the ways in which interfaces—internal and external—work and how internal capabilities affect the ways in which firms can acquire knowledge from outside.

What are the constraints on the speed of technological change in firms?

The study confirms that firms' existing capabilities determine the speed with which they can accommodate to technological change. This is of considerable policy significance, in view of the interest within the EU, as well as in other advanced countries, to develop policies which promote industrial change: the process of adaptation by industry to the new economic and technological realities.

Firms and economies are in a position to grow most quickly if they are present in the markets which, domestically and internationally, are expanding fastest. Most advanced nations—the EU member states included—have inherited industrial structures which in some ways are ill attuned to current economic conditions of demand and competition (OECD, 1992*c*). Besides the question of their presence on new, growing, and mainly high-technology markets, many EU countries have been faced in recent years with declining traditional industries, and serious economic consequences at a regional level. The grave unemployment throughout the EU has reinforced the interest of policy-makers in these issues. And a predominant question is the capacity for change, and the factors which affect it, at the level of the firm.

The study shows that the answers to these questions differ according to the characteristics of the economy, and in particular the presence of different types of firm—technology-intensive versus contingent, large versus small. But in all cases, public policy measures must be sensitive to the constraints on firms' freedom of action which are conferred by their existing technological and organizational competences. A sound understanding of these factors will allow the most effective measures for improving the relevant flows of technology and knowledge to be identified, as well as the prospects for adapting the technological specialization at national level.

The Organization of the Study

From the start, a broad definition of research and technology management has been adopted in our research study. The aim was not to describe the activities of specific functions or individuals, but to understand how managements relate to technological change. Technology management from that perspective encompasses aspects of management associated with the procurement of technology, with research, development, adaptation, and accommodation of technologies in the enterprise, and the exploitation of technologies for the production of goods and services. This includes product technology, process technology, and technologies serving management functions. For practical purposes, we will regularly use the shorter expression 'technology management' to refer to these activities throughout this book. In the next chapter, our understanding of technology management is further elaborated. The definitions and descriptions given there reflect our enriched understanding of the problems and methods of technology management at the end of our study.

The design of the study is illustrated in Figure 1.2. It was undertaken by a research consortium of nine European research institutes and universities. The consortium carried out interviews with top managers in over 250 enterprises located in nine European countries. All interviews were carried out on the basis of a common questionnaire and were complemented with desk research to bring together other relevant information in a systematic fashion. Altogether, ten separate reports were produced on the basis of these interviews, dealing with problems and patterns of technology management in four sectorally and six geographically defined environments (see Table 1.1).

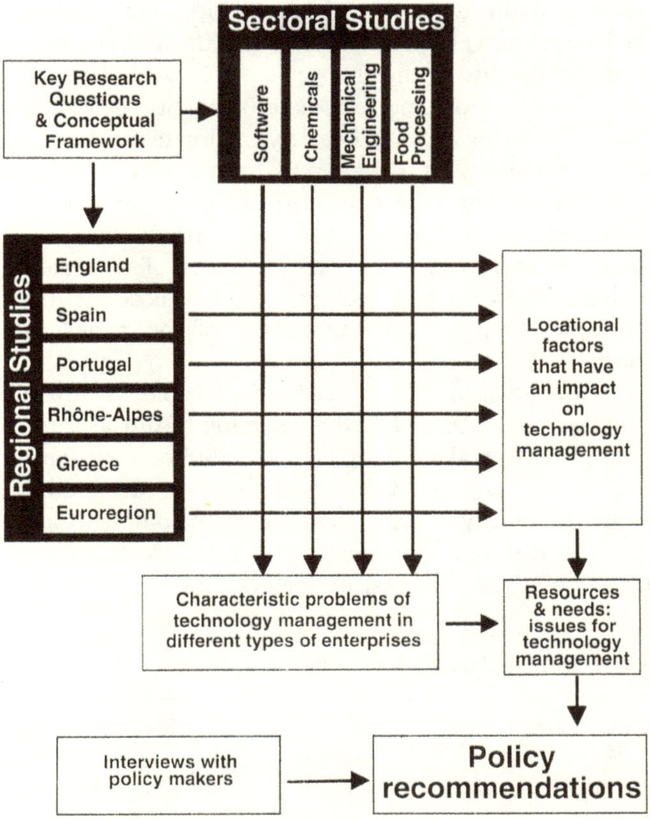

Fig. 1.2. The organization of the study

Although this may well have been one of the largest systematic surveys of technology management that has been carried out in Europe up to now, it could obviously not be comprehensive. However, by the use of a common methodology and by a careful selection of enterprises, every effort has been undertaken to establish a consistent and representative picture of technology management in Europe.

To make sense of the variety of problems and practices found in enterprises, the study was designed to address the principal underlying variables: the *industrial sector*, which defines the technologies, products, and processes of interest, the relevant industrial structures and the dynamics of competition; the *location*, which defines labour-market, infrastructure, and the administrative environment; and the *size* of the enterprise.

The choice of sectors and regions to be investigated was made on the basis of ensuring as wide a variety of circumstances as possible. The basic characteristics of sectors that were taken into account are their knowledge-base (technology-producing, science-based, or technology-using, traditional and mass production) and their industrial structure (the presence or absence of dominating large enterprises). The choice of regions took into account the level of economic development and the quality of the knowledge infrastructure.

The sectors selected for case studies (see Table 1.1) are internally far from homogeneous and further selection of sub-sectors was necessary in order to make the interviews in different enterprises comparable. Moreover, the 'regions' selected for case studies are in some cases extremely large, corresponding to complete countries. It should be noted that the aim was not to make comprehensive studies of countries, regions, or sectors, but to study technology

Table 1.1. Choice of sectors and regions for study

Sectoral studies	Regional studies
Software	Euroregion/Wallonie
Machinery	Rhône-Alpes
Chemicals	England
Food processing in less favoured regions	Spain
	Portugal
	Greece
Expert papers	
Electronics	
Aerospace	
Water	
Clothing in Ireland	
Chemicals	
Metals	
Construction	

management in a variety of economic and social-political environments.

The methodology and research results were reviewed at key decision points by a panel of industrialists and consultants. The members of this panel also contributed a series of expert papers on technology management in various sectors (cf. also Table 1.1). These reports were based on personal expertise and experience and did not make direct use of the conceptual framework developed for the study. Finally, a series of consultations was carried out in the different Members States of the European Community, involving interviews with representatives of governments, employers' associations, trade unions, and research institutions.

A typology of firms was constructed so as to ensure that inside every sector and region an adequate variety of firms would be approached for interviews. Thus, in every sectoral case study, care was taken to interview not just large enterprises, but also small firms; firms with a major R&D effort (a technology-intensive strategy) and firms with a minimal R&D effort (a technology-contingent strategy). In every regional case study, attention was paid not just to enterprises with a highly skilled workforce and a global approach to technology, but also to firms with an averagely skilled workforce, engaged in local sourcing of technology.[2]

The Organization of the Book

The structure of the book is as follows. Chapter 2 describes the findings of the study in overall terms. Chapters 3 to 5 present detailed analyses of technology-management challenges and firm responses in relation to three distinct types of firms: large firms, technology-intensive SMEs, and technology-contingent SMEs. In each case the analysis leads to specific policy recommendations, at regional, national, and Community level where appropriate. Chapter 6 provides a complementary policy perspective, and examines the role of the research and technology infrastructure with respect to technology management. A key dimension here is the distinction between 'core' and 'less favoured' regions of the EU, the

[2] In the study of Portugal, interviews were restricted mainly to new technology-based firms.

latter having been the site of very intensive infrastructural development in recent years. The chapter highlights a number of areas in which existing policy needs to be better adapted to local conditions, in particular to match the supply of research and technology infrastructure with the real needs of firms. Chapter 7 brings together the common themes from Chapters 3 to 6, and discusses the broader implications for policy and the role of public authorities.

2

Patterns of Technology Management in European Firms: *An Overview*

Ben Dankbaar

The translation of new knowledge into competitive products is one of the most important tasks of management. In fact, without much exaggeration it can be considered as the single most important task for the leadership of an enterprise. As the economist Joseph Schumpeter pointed out as early as 1912, the essence of capitalism is innovation, which he defined as making new combinations of markets, products, people, and technologies (Schumpeter, 1980). If the management of an enterprise does not succeed in making such new combinations or if it makes them too slowly in comparison with the competition, the performance of the enterprise will decline. The dynamics and growth of capitalist economies are determined by the decisions of enterprises to enter new markets, to introduce new products, to adopt new processes, new skills, and new forms of organization.

The use of new technologies has become an increasingly important part of the new combinations in our days. If technological knowledge is available and not used, the chances are great that competitors will take over. In view of the rapid development of technology and the countless opportunities for applications, enterprises need to be on the watch constantly. They need to monitor trends in relevant technologies, product and process developments in their markets, shifts in demand, the rise of alternative ways to fulfil the same needs as their present product, etc. They need to focus their own research and development activities in view of these trends, which requires a good understanding of the strategic orientation of the enterprise. Such monitoring and organizing of technology flows into and inside the company is what we call 'technology management'.

Technology management involves various activities, using numerous analytical tools and techniques to monitor technology and organize the implementation of strategic choices. At a more abstract level, we see this as involving three distinct competences, which we refer to as technological competence, entrepreneurial competence, and learning ability.

Technological competence: the ability to master the particular technologies that are relevant to the needs of the enterprise. Technological competence is basically knowledge and know-how. It is embodied in equipment, in patents as well as in human capital (tacit knowledge). It is created by research, development, and engineering activities, by licensing, purchasing, and training.

Entrepreneurial competence: the ability to generate and implement a strategy for the use and implementation of new technologies, coherently linked to an underlying corporate strategy, taking account of long-term trends in the evolution of technology, markets, and competition. The emphasis here is on the presence of a vision of the role of technology in the future competitive performance of the enterprise. Obviously, such visions differ in complexity and time horizon, depending on the size and technological basis of the enterprise.

Learning ability: the capacity to adapt organizationally and culturally in order to accommodate technological change. A firm's organizational structure and management routines determine its capacity to accommodate information on new technologies, market demands, and opportunities for innovation. In many cases, the effective exploitation of new technology also depends upon organizational adaptation.

It has been a guiding premiss of our investigations that innovation and technology are important to all enterprises in all sectors of the economy. Technology management, whether explicit or implicit, is therefore a relevant object of study, not just in large enterprises, but also in small ones; not just in high-tech industries, where research and development are standard activities, but also in traditional sectors, where technological change is often considered as an unwelcome threat to established products, procedures, and positions. Our study has focused on the problems that enterprises experience in the management of technology and the solutions they have found to overcome these problems. The final purpose of the investigation was to produce an information base—material, data, and insights—of practical use for policy development by decision-makers in the fields of science and technology and innovation policies. From a policy perspective it is important to acquire insights into the variety of technology-management practices in Europe. The appreciation of such variety must be the starting-point for any policy.

Our Methodology: In Search of Variety

We emphatically do not subscribe to the view that there is only one best way to manage technology. The challenges for technology management obviously differ between sectors. The characteristics of technologies and technological change in the semiconductor industry differ from those in chemicals or textiles. Different levels and different kinds of expertise are required in each industry. The industrial structure, technological content, and intensity of competition also differ between sectors, again requiring enterprises to develop specific levels and types of competence. That does not mean, however, that within any one sector or technological field there is one obvious 'best practice' that all enterprises should follow. In every sector, a variety of enterprises with different competences are able to survive. Nevertheless, the technological and structural characteristics of a sector define and limit the range of viable courses of action that an enterprise may choose.

The actual choices enterprises make depend on the competences they have already acquired (competence feeds competence), but also on the outside resources to which they have access. In that respect the geographical location of an enterprise may provide important opportunities, but it may also impose serious limitations. The presence or absence nearby of technologically qualified suppliers, the quality and accessibility of the knowledge infrastructure, the supply of skilled labour, the local regulatory regime: all these factors influence the feasibility of specific approaches for technology management in enterprises.

In order to uncover the full range of options, problems and possibilities for technology management in Europe, it is therefore of the utmost importance to ensure that management practices are investigated in a variety of sectors and geographical locations. The interaction between environmental variables (resources) and competences for technology management is clearly of relevance to policy-makers in this field. In some sectors or regions problems and deficiencies may arise with regard to technological competence, whereas in other cases organizational (entrepreneurial, learning) competences may be lacking. Policies in support of technology management would then have to be sensitive to these differences. The interactions between environment and competence may also offer distinct possibilities to influence the formation of competences indirectly, by effectuating changes in the environment.

Sectoral and regional characteristics narrow down the range of viable practices open to enterprises, but within that limited range enterprises may develop quite different competences. In any given sectoral or regional environment enterprises with quite different sets of competences will be able to survive. It is not just that different ways of operating may lead to the same result; the interaction between those different approaches contributes to the competitive performance of the whole sector or region. Our aim is therefore to map out the range of practices and the variety of competences that can be associated with competitive success in these different environments.

Since it was not possible to interview statistically representative samples of firms in the various sectors and regions selected for this study, it was important to ascertain that the characteristics of the firms chosen for investigation covered a broad spectrum, at least as regards the most important variables, so that they would reflect existing variety. A two-step approach was adopted, following the logic indicated above. First, four sectors and six regions were selected for case studies, with the objective of ensuring a diversity of regional and sectoral contexts for technology management. Then, individual firms were selected for examination on the basis of a fourfold 'sectoral typology of firms' (within the sectoral case studies) and a fourfold 'regional typology of firms' (within the regional case studies). Following this approach, the study was able to take account both of the general characteristics of regions and sectors and of the variety of firms operating within them.

Although we do not expect a single best practice to arise for each type of firm, we do expect variety within each type to be reduced. If that is indeed the case, it will be possible to identify cases where firms lack the competences they need in specific sectors or regions. It may also be possible to identify regions or sectors (at a regional or national level) that lack specific types of firms and therefore the competences that come with them, which could imply that something is missing in the industrial structure, possibly making that region or sector less competitive.

The Classification of Sectors

Our classification of sectors and our typology of firms within sectors is inspired by Pavitt's (1984) study of 'sectoral patterns of technical change'. Pavitt's analysis of the characteristics of about 2,000 innovations and innovating firms aimed to describe and explain similarities and differences amongst sectors in the sources, nature, and impact of innovations. A variety of qualitative and quantitative data were taken into account, and sectors were compared in terms of various factors:

The sectoral sources of technology used in a sector: whether it is generated from within or comes from the outside through the purchase of production equipment and materials ('technology-producing' versus 'technology-using');

The institutional sources and the nature of the technology produced in a sector: the relative importance of intra-mural and extra-mural sources of knowledge and of product and process innovations. This is related to the needs of the users, i.e. of the customers buying the products of the sector, and to the possibilities and means available for the appropriation of the benefits of innovative activities;

The characteristics of the innovating firms: their size and the intensity and direction of technological diversification.

The outcome of this analysis is the typology reproduced in Table 2.1. Pavitt distinguishes between 'supplier-dominated', 'science-based', and 'production-intensive' firms, the latter again divided up into 'scale-intensive firms' and 'specialized suppliers'. Pavitt's typology is a halfway house between a typology of sectors and a typology of firms. On the one hand, Pavitt is providing us with a typology of firms; on the other hand, he shows that the different types of firm within this typology tend to be associated with different sectors and associated technological trajectories. The lack of one-to-one correspondence between firms and sectors is a logical outcome of an analysis that starts with enterprise-level data. It reflects the fact that there is normally a considerable variety of firms within any sector. Below, we distinguish more clearly between

Table 2.1. Sectoral Technological Trajectories: Determinants, directions, and measured characteristics

		Determinants of technological trajectories			Technological trajectories	Measured characteristics			
Category of firm	Typical core sectors	Sources of technology	Type of user	Means of appropriation		Source of process technology	Relative balance between product and process innovation	Relative size of innovating firms	Intensity and direction of technological diversification
(1)	(2)	(3)	(4)	(5)	(6)	(7)	(8)	(9)	(10)
Supplier-dominated	Agriculture; housing; private services; traditional manufacture	Suppliers; research extension services; big users	Price-sensitive	Non-technical (e.g. trademarks, marketing, advertising, aesthetic design)	Cost-cutting	Suppliers	Process	Small	Low vertical

Production-intensive									
Scale-intensive	Bulk materials (steel, glass); assembly (consumer durables and and cars)	PE suppliers; R&D	Price sensitive	Process secrecy and know-how; technical lags; patents; dynamic learning economies; design know-how;	Cost-cutting (product design)	In-house; suppliers	Process	Large	High vertical
Specialized suppliers	Machinery; instruments	Design and development users	Performance sensitive	Knowledge of users; patents	Product design	In-house; customers	Product	Small	Low concentric
Science based	Electronics/electrical; chemicals	R&D Public science; PE	Mixed	R&D know-how; patents; process secrecy and know-how; dynamic learning economies	Mixed	In-house; suppliers	Mixed	Large	Low vertical High concentric

Note: PE = Production Engineering Department.

Source: Reprinted from Pavitt, K. 'Sectoral Patterns of Technological Change: Towards a Taxonomy and a Theory', 1984, 13: 343–73, with kind permission from Elsevier Science-NL, Sara Burgerhartstraat 25, 1055 KV Amsterdam, The Netherlands.

sectoral and firm characteristics, thereby further simplifying, but we hope also clarifying, the picture.

The classification of sectors we have adopted is based on two dimensions, indicating the intensity of competition and the character of the innovative processes in the sector. The *concentration ratio* is used here as an indicator of intensity of competition. This indicator has the advantage of simplicity, although we are aware of its limitations (competition may be intense, even if only a handful of firms dominate the market). It also plays an important role in the voluminous literature on Schumpeter's (1942) theses on the relationship between size and market structure, and the innovative performance of firms. We characterize the innovation process by distinguishing between product focus/technology-producing sectors on the one hand and process focus/technology-using sectors on the other. Obviously, by combining the focus of innovation (products or process) and the sources of technology (internal or external to the sector), we have eliminated two potential categories of sector (product focus/technology-using and process focus/technology-producing), which are not necessarily empty. The literature on product and technology life cycles, however, supports this approach. Generally speaking, most R&D is devoted to product development. In sectors with relatively high levels of R&D (technology-producing) product innovation usually predominates.

MATRIX I. *Typology of sectors*

		Innovation process	
		Product focus technology-producing	Process focus, technology-using
Concentration ratio	low	Specialized suppliers • software • mechanical engineering	Supplier-dominated • food-processing (in LFRs)
	high	Science-based • chemicals	Scale-intensive • food-processing (in advanced regions)

For mature products, the level of R&D is usually relatively low and the focus of innovation shifts towards process improvements which usually come from outside the sector (e.g. from suppliers of materials and manufacturing equipment).

Matrix I shows the sectors selected for our study according to the classification described above. Pavitt's fourfold typology also fits quite nicely into the same matrix. Note that sectors can move from one cell of the matrix to another in the course of their historical development. Indeed, the literature relating to the so-called industry life-cycle argues that in the course of time sectors will move from a product focus to a process focus and from low concentration to high concentration levels. The automobile industry is a case in point. In the early decades of this century, the industry was characterized by a large number of small manufacturers. Standardization of the product and process innovation involving scale economies led to a quick concentration process, which was already completed in the United States before the Second World War and in Europe lasted until the 1960s. Important product and process innovations now mainly come from suppliers of (electronic) components and production equipment. The food-processing sector was one of the sectors selected for special attention in our study, but only in the less-favoured regions (LFRs) of the European Union. In these regions, food processing is done mainly by small and medium-sized enterprises, using sometimes very old technology. The sector can therefore be characterized as 'supplier dominated' (low concentration, technology-using). In the advanced regions of Europe, on the contrary, food processing has become the domain of large enterprises. Although the latter are involved in considerable R&D on new products, innovation is fundamentally process-oriented. Food processing in the advanced regions can therefore be characterized as 'scale intensive' (high concentration, technology-using).

Within each of the sectors characterized in this way, one may find individual firms that do not seem to fit the sectoral profile. The prevailing technological regime may limit the range of viable options for an enterprise and the geographical environment may impose certain restrictions, but options are never reduced to the point where enterprises have no choices left at all. Indeed, the competitive and technological dynamics of a sector depend on the presence of a variety of firms and practices. For an understanding of the problems and prospects of technology management in the sector it is therefore necessary to investigate this variety. For that purpose, a typology of firms has been devised as illustrated in Matrix II. We are now 'zooming in', as it were, on one cell of Matrix I and looking at variety inside a given type of sector.

Matrix II distinguishes between small and large firms and between technology-intensive strategies with major R&D efforts (mainly the offensive and defensive strategies in Freeman's (1982) typology), and technology-contingent strategies with minimal R&D efforts (the imitative, dependent, traditional, and opportunist strategies in Freeman's typology).[1] The offensive and the defensive strategies have the highest technological content, but it is important to remember that other strategies are possible in relation to technology that may also lead to high or adequate competitive performance. Obviously, both these dimensions have different meanings in different sectors: a small firm in the chemical sector may be larger than a large firm in mechanical engineering and an imitative R&D strategy in chemicals may still involve a far larger outlay of R&D funds than what is necessary for an offensive strategy in the soft-

MATRIX II. *Sectoral typology of firms*

		Technology strategy	
		Intensive: major R&D effort	Contingent: minimal R&D effort
Size	small		
	large		

[1] Freeman (1982) uses scores on ten different dimensions of technology management (fundamental research, applied research, experimental development, design and engineering, production engineering and quality control, technical services, patents, scientific and technical information, education and training, long-range forecasting and product planning) to characterize six different technology strategies (offensive, defensive, initiative, dependent, traditional, opportunist). Freeman's characterization of strategies corresponds with the broad definition of technology management used here.

ware sector. The relevant indicators, in other words, will have to be relative measures at the sectoral level. Take for instance the chemical sector. On the basis of the sectoral characteristics (product focus, technology-producing, high concentration), one would expect most firms to be in the large/technology-intensive cell of the sectoral typology. But there are also small chemical firms and also both large and small firms with a technology-contingent strategy. For a clear understanding of the dynamics of technological change and technology management in the chemical industry, all four kinds of firms have to be investigated.

The Regional Typology

What specific characteristics of a region or country can influence the competence of firms to manage technological change? Is it possible to identify particular regional characteristics that favour—or conversely impede—the development of technology management in enterprises? Two 'regional' characteristics seem to be a priori relevant: the science and technology infrastructure and the situation on the labour-market (supply of skilled labour, wage levels, regulatory regime, etc.). Before we turn to these two aspects, it is important to point to some definitional difficulties when dealing with geographical dimensions.

In daily conversation, the word 'region' is reserved for geographical areas below the national level. If we speak of the regional environment of an enterprise, it is quite obvious that the size of the relevant region tends to increase with the size of the enterprise. For some enterprises the relevant region is a country, a continent, or even the entire world. The relevant size of the region also depends on the business function we are talking about. For technology sourcing, the boundaries are drawn wider than for finding skilled production personnel. Our approach to the management of technological change covers both activities. The study is dealing with technology management in a variety of regions, some of which are subnational, others national. These regions may be too small for some of the enterprises under investigation and too large for others. For the larger enterprises it will often be necessary to differentiate between the corporate level and the plant level (cf. Tödtling, 1990). The corporate level refers to the legal and taxable entity of a 'registered company', the plant is the local business entity, either identical with the enterprise (enterprise with only one plant) or a sub-unit of a larger enterprise (subsidiary or parent plant). Other organizational units between these levels (like business units) will also enter the picture. For each level, the relevant regional boundaries may be different.

Matrix III presents a typology of regions that takes account of the presence and strength of a knowledge infrastructure (what can popularly be called the 'Silicon Valley dimension') and the level of economic and industrial activity achieved in the region (the industrial experience effect, indicated for instance by the per capita level of income). The regions selected for this project have been placed inside the various cells of the matrix. The rather broad character of these 'regions' makes this a rather hazardous exercise. Obviously, each region contains sub-regions that would score differently from the way we have placed them in Matrix III. Nevertheless, their location in the matrix suggests that an appropriate variety of regions has been selected.

MATRIX III. *Typology of regions*

		Per capita income	
		Below EC average	Above EC average
Knowledge infrastructure	weak	Portugal Greece Spain	
	strong	UK	Euroregion Rhône-Alpes

Just as in the case of the sectoral typology of firms, our main interest is in developing a typology of firms that reflects the variety of experiences within regions. Inside a region with a weak knowledge infrastructure and relatively low per capita income, it is entirely possible to have some firms with a relatively well-paid, highly skilled workforce and a world-class technology base. As in

sectors, the dynamics of regional development can only be understood if the presence of all types of firm is taken into account.

The regional typology of firms presented in Matrix IV is based on the dimensions 'skill level of workforce' and 'technology sourcing'. Firms with an average or below average skilled workforce have been grouped into one category, as our main interest is in the distinction between firms with and without a highly skilled workforce. The average referred to here is the regional average and not the average for the sector to which the enterprise belongs, because we are interested in the dependence on and utilization of the human resources available in the region. Technology sourcing refers to the technological orientation of the firm: can it rely on local networking with similar firms and locally available bodies of knowledge, or does it have a global orientation, drawing its knowledge just as easily from abroad as from its immediate environment. Obviously, local networking for technology has a different meaning in the region Rhône-Alpes than in Portugal. The type of region in which a firm is operating is an important determinant of the options available to that firm. Nevertheless, given the region, firms still vary considerably in networking practices and skill levels.

MATRIX IV. *Regional typology of firms*

		Workforce skill level	
		Average and below	High
Technology sourcing	Local networking		
	Global networking		

The Main Influences on Technology Management Identified in the Study

The two factors which we found to have the most important influence on the demands faced by firms in the area of technology management are their size and the strategic position they have chosen towards technological change in their sector (shown by the presence or absence, size and type of R&D activities). Sectoral and geographical influences on technology management are less important; they can be explained mainly by the fact that a particular sector or geographical region involves higher or lower proportions of enterprises with these different characteristics.

This is a surprising result, since it challenges two widespread suppositions. The first of these is that a firm's commitment to technology is dictated by the industrial sector to which it belongs. For example, firms active in chemicals are considered high-technology firms; those active in foods are considered low-technology firms. In reality, sectors are not completely homogeneous and they allow firms to occupy a variety of niches. The problems faced by firms pursuing an 'offensive' technology strategy have more in common with those of enterprises with a similar strategy in other sectors than those following strategies which are less oriented towards technological leadership in their own sector.

The second assumption challenged by our results is that an offensive technology strategy is only available to companies in better developed regions of Europe, since these strategies depend upon good connections with a high-quality local knowledge infrastructure. In fact, the problems of small, traditional firms in less-favoured regions (LFRs)—commonly regarded as the type of firm suffering from lack of infrastructure—appear very similar to such firms in better-endowed regions. Moreover, the existence of some dynamic, technology-based firms in LFRs illustrates the increasing capacity for such enterprises to develop networks on an international scale, despite the constraints of their local environment.

The problems of firms from different sectors that are similar in relative size and technology strategy are often more alike than those of different types of firms operating in the same sector. Obviously, in a science-based industry, the R&D activities of a large, technology-intensive enterprise will probably be larger and maybe more scientific and high-tech than those of a similar type of enterprise in a supplier-dominated sector. We suggest, however, that the attitudes, problems, and needs of these enterprises are often more similar than those of other types of enterprises in the same industry. Thus, we find it useful to distinguish four types of enterprises that can be found in almost all sectors and that each have their distinctive practices and problems: technology-intensive SMEs, the large technology-intensive enterprises, the large technology-contingent enterprises, and the technology-contingent SMEs.

These four types of enterprises can be found in practically every sector, but in every sector some types are more common than others. In fact, we would suggest that these types are the characteristic types of enterprises in each of the different types of sector distinguished above. In each of the four types of sectors we find each of the four types of enterprises, but within each type of sector one type of enterprise is dominant. This is illustrated in Matrix V, which combines the two sector-oriented typologies. The darkened cells in the matrix show where the characteristic type of enterprise can be found within each type of sector.

Below follow some comments on each type of enterprise, characterizing their basic problems of technology management.

MATRIX V. *Four characteristic types*

		Technology-producing		Technology-using	
		Technology-intensive	Technology-contingent	Technology-intensive	Technology-contingent
Low concentration	small	**small RD+ specialist supplier (SS)**	small RD– SS	small RD+ supplier-dominated (SD)	**small RD– SD**
	large	large RD+ SS	large RD– SS	large RD+ SD	large RD– SD
High concentration	small	small RD+ science based (SB)	small RD– SB	small RD+ scale intensive (SI)	small RD– SI
	large	**large RD+ SB**	large RD– SB	large RD+ SI	**large RD– SI**

Key: RD+ = above average R&D intensity
RD– = below average R&D intensity
SS = specialist supplier
SD = supplier-dominated
SB = science-based
SI = scale-intensive.

Small Technology-Intensive Enterprises

The literature on small, technology-based enterprises sometimes seems to suggest that such firms have no problems of technology management. At first sight, there is no lack of technological competence in most of the small science-based or technology-intensive enterprises. In fact, they are practically defined by this competence. Moreover, these firms are supposed to be entrepreneurial, flexible, and eager to learn more about their technologies. These firms are open to communicate with customers and often concentrate on a limited number of sophisticated customers. In our research, the prototype of these firms has been provided especially by small and medium-sized mechanical engineering firms and software firms.

These firms themselves often recognize that the main factor defining their strength is at the same time also a potential weakness. The fact that they have specialist knowledge, often tacit knowledge in the heads and hands of a few key persons, is the basis for their competitive performance. By concentrating on a limited area of knowledge, however, they run the risk of neglecting the impact that developments in other technological fields may have on their own business. This problem has surfaced quite often recently where new developments are based on the fusion of two or more existing technologies, only one of which is usually available in a single small firm. Apart from the lack of technological competence, lack of finance can also become a problem. Investment needs multiply, if new technologies are added. Given the small size of these enterprises, risks cannot easily be balanced by other activities and banks will be reluctant to put up the necessary means.

Frequently proposed solutions to these problems are co-operation with other technology-based firms (in research, but also in the development and marketing of complete packages of products with different technological content); the cultivation of links to universities and research institutions; or the establishment of ties to larger enterprises that can serve as antenna for signals on technological changes, but also as a source of (financial) support if new technologies have to be acquired rapidly in order to maintain competitiveness.

Small technology-intensive enterprises are aware of these solutions, but we find that they often have strong reservations about co-operation with other enterprises, no matter if the partners are large or small. Independence seems to be considered as a value in itself that is guarded jealously. Co-operation with universities offers less problems in that respect, but is of little use if the bottleneck is investment in equipment for the application of new technologies. Contrary to the situation in Japan, small manufacturing enterprises in Europe usually do not belong to a kind of family of firms related to a larger enterprise (*keiretsu*), on which they can rely for technical and financial and management support.

The alternative to co-operation is internally generated growth, which must be based on continuous investment in human resources. The acquisition and expansion of technological competence in these enterprises is, even more than elsewhere, a question of human resources management. Technology management in these enterprises is hiring, training, and retaining the right people. From a labour-market perspective, technology-intensive SMEs can only compete with large science-based enterprises, if they offer scientists and engineers a more attractive environment for the development and implementation of new technologies. An important precondition for the further success of small, technology-based firms will be a large and continuous supply of qualified engineers and technicians.

Technology management in large, technology-intensive enterprises may lead to the creation of additional small, technology-based enterprises (spin-off firms), if the decision to concentrate on specific core technologies leads to the elimination of promising projects that fall outside the core. In the recent past, many large firms (science-based as well as mass production) have also become the source of new small consultancy firms, as a result of the efforts to increase flexibility and reduce costs by the elimination and subsequent outsourcing of staff and support activities. The experiences learnt there could possibly be used fruitfully for the creation of new technology-based manufacturing firms as well.

Large Technology-Intensive Enterprises

Large technology-intensive enterprises compete on the basis of technological competence. They are leading innovators in their sectors or follow the leaders closely. In all sectors, the largest technology-intensive enterprises tend to adopt a global perspective. Competitive strategies aim at technological breakthroughs, followed up by rapid product innovation. Secrecy and patenting are important instruments to protect the investments they make in research. Anything that shortens the time between product definition and full-scale production confers competitive advantage and may even allow enterprises to catch up with early innovators. The prototype of large, technology-intensive enterprises has been provided in our study by the large chemical (especially the pharmaceutical) enterprises and the large electronics multinationals.

The awareness of science and technology as management problems is obviously greater in these enterprises than elsewhere and the need for some form of 'technology management' is widely conceded. Quite different opinions exist, however, as to what this really means. There is still a strong tendency to counter problems of competitiveness with the solution 'more research' and to think of technology management in terms of R&D management. In spite of that, most of our respondents would agree that—with few exceptions—the challenges facing large technology-intensive enterprises in Europe are not to be found in a lack of technological competence. Spending more on research may be necessary in some cases, but it is doubtful that this will improve competitiveness if nothing is done to improve the efficiency of the organizations in terms of the uses that are made of research. Bureaucratic organizations with functionally specialized departments that resist attacks on their autonomy and lack incentives for co-operation, have hampered entrepreneurship and slowed down initiatives for change.

Technology management in science-based and technology-intensive enterprises is therefore not so much a question of attracting additional technological competence, but of organizational redesign. The introduction of new organizational structures, however, is far from simple. Even in firms where the interaction between organizational change and technological change is clearly appreciated by top management, the translation into practical behaviour is often limited. The need for technology management and organizational redesign may be well understood and the (textbook) solutions known, but the difficulties and costs of implementing more focused and at the same time more flexible organizational forms, are systematically underestimated. As a result, board-level willingness to invest resources in the creation and support of new organizational structures often remains insufficient.

In the field of human resources management, some management development policies are in place in relation to research personnel. These are based on the traditional organizational structures and will be affected by organizational redesign, since this will involve de-emphasizing functional specializations and the creation of multidisciplinary teams and departments. Especially the need to combine technological and scientific knowledge with entrepreneurial competences requires a closer interaction between research and general management functions and where possible also switching between different roles during the career of a scientist or engineer.

What becomes clear from our interviews is that large science-based firms generally do not look at universities for advanced research results. Apart from a few special world-class institutes, universities are basically considered as a source of well-qualified personnel. To the large enterprises, research at universities is mainly justified if it is needed to provide an adequate teaching environment. This role of universities as teaching institutions, however, cannot be strongly enough underlined. Whether research laboratories stay where they are or move somewhere else depends very much on the continuing supply of qualified personnel.

Large Technology-Contingent Enterprises

Large, technology-contingent enterprises may possess considerable technological competences and employ large numbers of scientists and (above all) engineers. Technological change and product innovation are not at the core of their business strategy, however. Depending on the sector in which they are operating, these enterprises can survive at a distance from the technological frontier on the basis of other competitive advantages, conferred by design, scale, low costs of inputs, or effective organization. In our project the prototype of the large technology-contingent enterprise was provided by large steel and aluminium manufacturers (treated in an expert paper). Large technology-contingent enterprises, it should be underlined again, can also be found in other sectors than the scale-intensive sectors. In food processing, for instance, we have encountered large enterprises that employ almost no technicians, let alone having R&D activity.

Large, technology-contingent enterprises have to defend their size, i.e. their market share. Unable to do so on the basis of radical innovations, they have to make sure that they deliver what the customers want, be it low prices, high variety, high quality, unique features, special delivery service, special packaging, high speed delivery, or any combination of these. With relatively mature product technology and large-scale production processes, the issues of technology management in these enterprises are quite different from those in the large, technology-intensive enterprises. In technology-intensive enterprises, the basic problem is the transformation of research results into competitive products, i.e. the problem is one of innovation management; in technology-contingent enterprises, the basic problem is to utilize available technologies and to organize a sometimes complex system of suppliers and sub-suppliers to serve the needs of customers.

Large, technology-contingent enterprises are often operating in mature markets. Although the basic product and process technologies in these markets are considered mature, incremental change may still play an important role and may require continuing investment in research and development activities. The senior technical officers in these enterprises have a natural tendency, based on their expertise, to concentrate on these incremental developments in product and process technology. However, the competitive performance of these enterprises is equally or even more dependent on effective marketing, attractive design, speed, flexibility, and external networking. This is to a large extent a question of organizational design and not of technology. The competitiveness of well-designed organizations, however, could probably be further enhanced by more effective use of information technology in various supportive and indirect functions. Although considerable investments have been made in information systems for logistics, marketing, order processing, and production planning, their impact has often been disappointing. The individual systems have been optimized for individual functions in traditional organizations. What is needed is an integral approach to the use of information and communication technologies, based on an integral 'systemic' view of the internal as well as the external organization of the enterprise.

All this also poses considerable challenges in the area of human resources management. In scale-intensive branches like steel, process automation has already resulted in the creation of huge manufacturing complexes employing only a few highly skilled workers. Productivity increases in assembly operations, for instance in the automobile industry, will also lead to the disappearance of thousands of jobs. Some of these will re-appear in the supplying industries because of increased outsourcing by the assemblers. Both there and in the large mass-production operations, skill requirements will be increased by the introduction of new technologies as well as new organizational structures that ensure higher flexibility and productivity.

Small Technology-Contingent Firms

Whereas large technology-contingent firms may still have considerable technological competence, small technology-contingent firms are usually lacking in this respect. They may have some competence (know-how) in a limited area, but they *lack the capacity to investigate and assess new developments* in that area, let alone developments in other areas that may threaten their competitive position. Especially in technology-using sectors, the difference between large and small technology-contingent firms may be impressive. In our research, the study of food processing in the less favoured regions of Europe has provided the prototype of such enterprises.

It should be noted that in the past the small technology-contingent firms often did not need a capacity for technology management, because they were operating in a rather stable technological trajectory. They could survive on a reactive strategy that required minimal investment in technological capacities, depending on the sector they were in. Changes in technology were mainly process changes (e.g. the introduction of NC-machines or PLCs for process control) that required some adaptations, but no active management of technological change. Currently, demands upon these firms are rapidly increasing, in terms of technological competence as well as in terms of organizational abilities to ensure constant high quality and reliability. Competition is coming mainly from other European firms and is imposed by increasing demands from large customers that are streamlining their supply chains.

An important bottleneck in these firms is the fact that the responsible management lacks time and often also the inclination to adopt a long-term perspective on the firm and its technology. This is especially the case in small family-owned businesses, where the founder-proprietor has not found a suitable successor—or has been unwilling to search for one.

The lack of an adequate marketing function is often quoted by these firms as an important bottleneck in their business. The link to technology is more direct than it may seem at first sight. Quite often, these firms get access to new technologies through their customers (which may be large firms). More generally, it can be argued that the introduction of new customers and new markets would have a profound impact on most of these firms, which often depend on a handful of traditional customers. The introduction of an active marketing function would therefore introduce new demands upon their efficiency and technological competence. This could be especially true for efforts to sell on foreign markets. In fact, the marketing function refers here not so much to the ability to sell things, but to the ability to communicate with (new and potential) customers and to translate customer needs into new and improved products.

In this category of small firms, the existing technology programs of the Community are widely considered unattractive. These firms would be more interested—if at all—to co-operate with other firms from the same country or from a specific supply chain than with a number of foreign enterprises. Moreover, technology would quite often not be the focus of such co-operative activities. Better marketing and easier access to finance could be other reasons for co-operation.

Locational Factors that have an Impact on Technology Management

In all regions, competences for technology management differ widely between enterprises. Consequently, these enterprises also have different problems. We find that very often these problems are defined more by their size and technological competence (i.e. if they have chosen a technology-intensive or a technology-contingent course) than by their location. A key issue common to all regions, independent of the existing knowledge infrastructure and per capita income, is the need to improve the sensitivity of traditional SMEs to technology management. Their predominantly inward-looking, short-term vision calls for an active policy and specialized support in order to increase their tendency to adopt an explicit technology strategy and, if necessary, not to hesitate to go into global networking. A particular issue for them is the improvement of supply chains.

Summarizing our findings on the locational perspective on technology management, two points can be made. Firstly, locational factors determine to a considerable extent the resources that enterprises have at their immediate disposal. This obviously has an impact on the problems that enterprises encounter in their efforts to manage technological change. Lack of skilled labour, high costs of finance, or the absence of research laboratories can present important bottlenecks. However, secondly, in every location in Europe, even in the less favoured regions, there are enterprises that have found ways to overcome the locational limitations and access resources and knowledge that are available in other regions and countries. The location, in other words, can be considered as an important part of the resource base of the enterprise, but this resource base is not the sole determinant of technology management capacity. Consequently, not all enterprises in a region can be expected to experience the same problems and bottlenecks in technology management. In that sense, there does not seem to be a clear basis for policies in support of technology management in enterprises that take a regional perspective without considering the characteristics of the firms that are being targeted. For instance, if policies in support of technology management in the less favoured regions were designed on the assumption that all potential users are characterized by low skills and local sourcing of technology, they may fail to be useful for the firms that may actually hold a greater promise of success: firms that have already managed to build up a skilled workforce and international links to sources of technology.

Three major locational factors have been identified that influence the resources available for technology management in enterprises:

1. The supply of qualified labour, i.e. the educational and training system. Here we have to distinguish between the supply of skilled labour and the supply of scientists and engineers.
2. The supply of technical knowledge, where we have to distinguish between technology transfer institutions, supplying existing technologies, and advanced research laboratories (inside or outside universities), supplying new technologies.
3. The prevailing structure of the economy, where we have to distinguish between the existence of networking relations with other similar firms (e.g. the so-called Local Production Communities and Industrial Districts) and the presence of supply-chain type relations, focused (usually) on one or a few large dominating enterprises.

The relevance of the first two of these types of resources has been generally recognized and considerable efforts have been undertaken by regional and national authorities as well as through European Community initiatives to strengthen the knowledge infrastructure and to upgrade the educational system and institutions for vocational and further training, especially with respect to the needs of SMEs. The third locational factor appears to be much more important for technology management than is often recognized. Enterprises learn most and most quickly from other enterprises. The presence or absence of potential supply-chain networks and the presence or absence of horizontal networking amongst specialist competitors (Local Production Community/ Industrial District), appears to be of much greater relevance than the presence of a public technology transfer infrastructure, which has received so much attention in (regional) RTD policy-making.

Key Issues of Technology Management

Table 2.2 combines some of the basic findings of our sectoral and regional studies. On the left-hand side of the table, starting from the basic competences of technology management that were identified in the introduction to this chapter, it is indicated for each type of enterprise whether the necessary competences are likely to be present or not, and where important needs for improvement can be identified. The right-hand side of the table indicates the uses that each of these types makes (or should make) of the resources that were identified above in our discussion of the regional environment.

Two general observations need to be repeated and kept in mind while reading this table. First, in every sector and in every region, enterprises can be found that come close to each of these types and consequently have the specific capabilities, shortcomings, and problems related to that type. Second, a corollary observation is that a seemingly increasing number of enterprises can overcome the limitations imposed by their geographical environment and acquire the resources (especially knowledge) they need from other places.

Considering the capabilities and shortcomings of different types of enterprises, the following points need emphasis:

1. *The predominant problem that technology-intensive SMEs are faced with is the need to develop an explicit technology strategy that recognizes the limitations imposed by specialization.* This will involve the gathering of specific information on long-term technological

trends and the determination of the core technological competence of the enterprise in relation to these trends. The ability to do so may not be available. Co-operation with other non-competing specialists may be needed, both to develop a strategic vision and to achieve a capacity for the production of complete systems incorporating the technologies of the participating firms. Alternatively, large customers may serve as an important source of information on related technology trends and future needs.

2. *Large technology-intensive (science-based) enterprises are confronted with the need to redesign their organizations for more effective management of innovation processes.* At the most general level this means to introduce better communications both inside the enterprise and with its technological environment. This will involve the introduction of multidisciplinary project teams and a breakdown of functional specialization structures. R&D activities will be more integrated with other business functions: they need to remain market-driven, but a long-term orientation is necessary to prevent short-termism and incoherence ('third generation R&D management'). Specialization on core technologies needs to be accompanied by a greater willingness to buy technology from outside. The capabilities to carry out technology surveys and develop technology strategies are basically in place, although they are not always used to that effect.

3. *Technology-contingent (traditional) SMEs also need to develop a longer-term orientation towards their technological environment.* The capacity to do so is not available and for the time being the highest priority of these enterprises is to increase their in-house technological competence. The main means to do so are the employment of engineers (and giving them time to think about the future of the business), training of the workforce, and management training. Information on relevant technologies and trends is usually available from branch organizations and technology transfer institutions. The main problem is not accessibility of information but the reluctance of enterprises and their owner-managers to face the need for managing technological change in the enterprise before it is too late.

4. *The main problem for large technology-contingent and mass production enterprises is their lack of organizational flexibility.* Their competitive performance depends less on the ability to be first with a specific technology, but more on the ability to introduce minor improvements and achieve continuously improving levels of quality

Table 2.2. Technology management in enterprises: needs versus competences and locational resources

Type of Enterprise	Competences			Locational Resources					
	Technological competence	Entrepreneurial competence	Learning ability	Public technology transfer infrastructure	High-tech/ scientific infrastructure	Vocational and continuous training system	Scientific and engineering educational system	Specialized regional economic structure	Presence of large and/or multinational enterprise in the region
Technology-intensive SMEs	Competent in a limited area; need to acquire broader range of techno-logies more specialist information on techno-logical envir-onment needed	No urgent needs for organizational redesign	More emphasis for linkages with outside related expertise needed	Usually not specialized enough for technology-intensive firms	Needed; difficult communica-tion	Less important	Highly important as a source of qualified personnel	Important potential source of competence	Less important; potential source of knowledge if customer

Large science-based enterprises	Competence available	Need organizational redesign (defunctionalization) to support innovation management	Capacity (training, R&D management) usually in place; need for greater external orientation	Not needed	Only centres of excellence really important	Less important	Highly important, but also access to international supply	Less relevant	Competition for qualified staff; potential stimulus if in same sector
Traditional SMEs	Insufficient competence; often also insufficient analysis and evaluation of technological options; information needs relatively simple; sources available and accessible	No priority; often need for more marketing skills	Need for technical managerial staff; training of workforce	Needed, but not very often used	Only indirect links, if at all	Highly important	Less important	Important potential source of competence	Opportunities for learning in supply chains
Large mass production enterprises	Competence available, but often not used	Need organizational redesign to support continuous improvement; supply-chain management	More emphasis on supportive functions and technologies (information systems)	Not needed	(Limited) use as source of technology and long-term research	Important, but if necessary complemented by internal training	Important as source of qualified personnel	Potential benefit of specialized suppliers	Competition on labour-market; potential stimulus if in same sector

and productivity. Research, development and engineering functions need to take incremental improvement activities more seriously. This requires a redesign of their organizations for more efficient management of the total production system, improving communications between engineering, production, distribution, and marketing functions. In the past, these enterprises have often introduced information technology in the expectation that it would improve the flexibility and effectiveness of their organizations. It has become clear that the sequence is different: once the organizations have been redesigned for flexibility, information technology can be introduced to support the new organizational structures. Special attention also needs to be given to the role of technology and technology transfer in relations with suppliers (supply-chain management).

Considering the supply of resources for technology management, the following points can be made:

1. *Within the knowledge infrastructure, it is important to differentiate between the system for education and training and the system for research, development, and technology transfer*. The importance of the system for training and education is undisputed. The relevance of public (university) research for technology management in enterprises, however, is less clear than is often assumed. Although there are important exceptions, university research is usually not considered nor expected to be of immediate relevance for the improvement of technological competence in enterprises. Institutions and resources for technology transfer are often not used by the enterprises that are supposed to need them and for whom they have been made available. That does not make technology transfer institutions superfluous, but it does seem to imply that additional means for technology transfer are unnecessary if demand for technology especially in SMEs does not increase.

2. *Enterprises learn most from other enterprises, be it as customers or as suppliers*. The importance of regional economic specialization and interaction (industrial districts) and of supply-chain networking for technological learning is undisputed, especially to SMEs. In that respect, large enterprises can be considered as important technology transfer institutions (to their suppliers). Their potential role for regional development from this particular perspective seems to have been underestimated. The fact that enterprises learn most and most quickly from other enterprises has so far not been used effectively in the organization of technology transfer programmes.

3. *In the case where several large firms of the same sector are located in the same region, competition is an important stimulus for the development of competences for technology management.* Competition between large enterprises on the labour-market may lead to rising labour costs, but also to higher demands on the educational system in the region.

4. *Almost by definition, the supply of knowledge and other resources (risk capital, skilled labour) is lower in the less favoured regions than elsewhere in Europe.* In recent years, a considerable technology transfer infrastructure has been installed in these regions. The majority of enterprises there, however, are traditional SMEs, which still have to be motivated to use the available technology transfer institutions. In that respect they do not differ very much from traditional SMEs in other regions. There is, however, a small but growing number of enterprises in the less favoured regions that have found it possible to overcome the locational limitations and develop considerable technological and entrepreneurial competence. These more sophisticated enterprises are in fact experiencing some other disadvantages that are highly specific to their location: lack of sophisticated customers; negative country image; and the 'invented elsewhere syndrome'.

Building up Competences for Technology Management

As noted above, the capacity for technology management involves in our view basically three competences: technological competence, entrepreneurial competence, and learning ability. These three competences are considered as cumulative: entrepreneurial competence makes use of technological competence and the ability to learn allows for the absorption of new knowledge and the renewal of entrepreneurship. The 'learning organization' has the capacity to manage technological change in the enterprise as a process of continuous improvement, innovation, and organizational change.

Building up the capacity for technology management will be different in different enterprises. The obvious starting-point is the existing competences. The four types of enterprises that were distinguished and described above typically have different competences and therefore different starting-points in the process of building up the capacity for technology management. The shortages and bottlenecks they face are different, and the resources they can command are likewise different. Different enterprises therefore have a different road to travel, but the aim is the same for all of them: to learn how to manage technological change and create a solid basis for their competitive performance.

Acquiring the capacity for technology management is a process of organizational development. This means that activities, functions, rules, and organizational routines have to be developed that allow the enterprise to acquire and implement technological knowledge quickly, cheaply, and competitively. Only for the traditional SMEs is the acquisition of technological competence a first priority. For all other types of enterprises (large enterprises and technology-intensive SMEs), the major bottleneck is not technological knowledge but organizational competence. These are the enterprises that operate most frequently on international markets. Thus, the competitiveness of European industries depends on the performance of precisely the types of enterprises for whom technological competence is not a major bottleneck.

Our confrontation of the needs of the various types of enterprises with the location-bound resources brings us to the following conclusions. For firms that want to acquire technological competence, a technology transfer infrastructure is basically available everywhere in Europe. For firms that want to acquire organizational competences, no comparable public infrastructure for the transfer of organizational knowledge and experience is available. RTD programmes and scientific and technical education at universities or polytechnics pay little or no attention to related problems of organizational change and redesign. There is, of course, an infrastructure of private consultants for organizational change, the use of which can be and is increasingly encouraged. Our policy recommendations therefore concentrate on the development, acquisition, and transfer of organizational competences. The finding that enterprises learn most from other enterprises needs to be incorporated more fully into programmes for organizational learning that are available to enterprises.

It should be noted that the most important factor is that enterprises feel the need to acquire these competences and act out of the urge to survive in competition. In that respect, our discussion has shown that any policy in support of technology management will have to be part of a larger policy framework that takes account of all aspects of industrial dynamics.

3

Large Enterprises: *Policies to Promote Best Practice in Cross-Functional Technology Management*

Rod Coombs

Introduction

The Nature and Significance of Technology Management in Large Firms

It is widely accepted that technology plays an important role in competitive relationships between firms, and between regions and nations. Much of the academic research and policy analysis which has examined and elucidated this problem over recent decades has been concerned with the production, volumes, and flows of technology itself. However, it is increasingly becoming clear that, at least in the relationships between advanced economies, and in the competitive battles between firms within such economies, it is often not the technology itself which determines outcomes but the management of technologies. To put it simply, most large firms do not suffer principally from a lack of technology, but from a lack of the ability to mobilize it quickly and effectively enough to achieve competitive advantage. Of course, there can be a wide range of reasons for such failures. However, some at least, of the reasons for large firms failing to create competitive advantage out of technology are due to deficits in a specific set of procedures and skills which we can identify as technology management skills. These are the focus of this chapter. The first task is therefore to briefly define these skills and activities.

Technology management is an activity whose scope is still not completely defined. Although some features of the activity are very clear, others are still subject to debate and differing emphasis in different contexts. Rather than try to give a dogmatic definition, it is more productive to identify some major themes which define the territory which technology management can cover.[1] In so doing we focus in particular on five aspects which are of particular significance for large firms.

1. *The management of R&D*; but with specific emphasis on integration of the R&D function's goals with those of the business unit(s) which it serves. In other words, R&D management as part of a managed process of technological innovation. In recent years large firms have also had to contend with the difficulties of co-ordinating the activities of multiple R&D centres dispersed across countries and providing differing types of skills and resources.

2. *The management of external technology flows into the enterprise*; this involves external sourcing of technology; collaborative linkages; and relationships with public R&D activities.

3. *Analysing the long-term technology issues relevant to the strategy of the enterprise*; this involves mapping the technological competences within the business(es) and within their R&D units; taking responsibility for husbanding and developing appropriate technology competences; and ensuring that these competences are properly considered in the business and corporate strategy agendas of the firm. This involves balancing the demands on R&D for shorter-term product-oriented work and longer-term knowledge-building work.

4. *The recognition and exploitation of the new-business creation potential of R&D outputs*; and ensuring that corporate strategists take appropriate account of these possibilities in shaping the future of the firm.

5. *Championing the effective exploitation of technology throughout the enterprise.* This can take two major forms. The first is the use of organizational measures to optimize the new product development and launch process beyond the R&D pipeline and across all the other functions of the enterprise (especially manufacturing). The second is the internal use of technology (especially information technology) to enhance the speed and effectiveness with which the enterprise continuously improves its products and services.

[1] For a variety of definitions the reader can consult Dussauge *et al.*, 1992; Roussel *et al.*, 1991; and Dodgson and Rothwell, 1994.

The first four of these points could be seen as a 'top-down' perspective on the enterprise which is concerned with strategic positioning by means of analysis of core competences (Prahalad and Hamel, 1990); and with utilizing technology in the products and services of the enterprise to gain competitive advantage. The fifth point is a more 'bottom-up' perspective which is concerned with the optimization of the contribution of technology to the operational management of the enterprise. The issues raised under this theme connect to the separate but related agendas of such topics as 'lean manufacturing' and business process re-engineering.

These points can be summarized by saying that technology management is an activity which arises from the gradual convergence and overlap of three agendas; the innovation agenda, the manufacturing agenda, and the strategy agenda. Its origins lie in the gradual recognition over recent decades that innovation involves making R&D and technology become market-facing; that manufacturing is about flexibility as well as efficiency; and that strategy is about long-term accumulation of competences as well as choosing particular competitive stances for particular markets.

Technology management is therefore concerned with configuring an organization so that it maintains an innovative potential over long periods, not simply innovating sporadically in an *ad hoc* manner. This involves, in the language used in the study, having the appropriate technological competences, the appropriate organizational competences (learning ability), and the appropriate entrepreneurial competences.

The *locus classicus* of the technology-management practices described above is in the large technology-intensive firms which are a prominent part of the firm population in the 'science-based sectors'. Not all firms in the science-based sectors, however, are large, technology-intensive enterprises and in the final analysis, the characteristics and problems of technology management in enterprises depend more on the type of firm they are than on the type of sector they operate in. In fact an important finding of the research was that the two most important characteristics of enterprises which influence the demands they face in the area of technology management are their size and the strategic position they have chosen towards technological change in their sector. In this chapter we are concerned with two types of firm: large firms with an 'active' technology posture (the 'technology-intensive' firms), and large firms with a 'reactive' posture towards technology (the 'technology-contingent' firms).

Technology Management Issues in Large Enterprises

The General Situation

Table 3.1, which reproduces part of Table 2.2, is a useful starting-point for considering the current situation with respect to technology management in large European firms. It summarizes the empirical findings of the study, across sectors and countries, with respect to the capabilities of the two types of large firm in a number of broad aspects of technology management.

If we focus first on the technology-intensive large firms, we see that the study finds them to be characterized on average as reasonably capable in strategic analysis of technology, and also in the protecting and surveillance of technology; but lacking some capabilities in organizational redesign to support innovation; and needing to increase external technology networking. In addition the table shows that such firms implicitly depend on the local scientific and technical infrastructure, but principally as a supply of manpower rather than as a supply of specific research results. To this we must add the fact that, although some large technology-intensive firms do have substantial experience and expertise in what we have called 'technology mapping', this is not absolutely universal. There are shortfalls in some areas, and some firms still have their capability fragmented across different business divisions rather than firmly co-ordinated from a central corporate point.[2] This is a function of the particularities of corporate histories, levels of diversification, management styles, etc.

The overall pattern which emerges from the findings is that these large technology-intensive firms, although strong in many aspects of technology management, do have specific deficits. Specifically,

[2] The role of a corporate technology management centre can be described as follows: to analyse the structure of the overall technology portfolio; to ensure that a technological competence in one business is known to and available to other potential user business in the group; to identify technical competences which straddle businesses and to take steps to strengthen them through 'horizontal' organizational links and through small special budgets; and to consider the overall technology portfolio and inject an appreciation of this portfolio into the broader strategic management processes of the company.

Table 3.1. The capabilities of two types of large firm in technology management

Capabilities for technology management and locational resources	Type of enterprise	
	Large science-based enterprises	Large mass-production enterprises
Strategic analysis and evaluation of technological competence	Capacity available, but not always adequate and not always used	Limited capacity available, and often not used
Innovation management and organizational redesign	Need organizational redesign to support innovation management	Need organizational redesign to support continuous improvement; supply-chain management
Learning, networking, and R&D management	Capacity in place; need for greater external orientation	More emphasis on supportive functions and technologies (information systems)
Protection and surveillance	Capacity available	Capacity available
Public technology transfer infrastructure	Not needed	Not needed
High tech/scientific infrastructure	Only centres of excellence really important	(Limited) use as source of technology and long-term research
Vocational and continuous training system	Less important	Important, but if necessary complemented by internal training
Scientific and engineering educational system	Highly important, but also access to international supply	Important as source of qualified personnel
Specialized regional economic structure	Less relevant	Potential benefit of specialized suppliers
Presence of large and/or multinational enterprises in the region	Competition for qualified staff; potential stimulus if in same sector	Competition on labour market; potential stimulus if in same sector

they need to address and improve their technology management practices in three major areas: cross-functional management for innovation; strategic technology mapping; and technology networking and outsourcing.

The situation is somewhat different for the large technology-contingent enterprises (many of which are mass-production enterprises). It is clear from the empirical findings of the study that these companies have all of the needs of the large technology-intensive firms, but they have some additional needs as well. These are:

1. *Organizational redesign* to achieve tighter integration of the supply chain and continuous improvement;
2. *More intensive use of information systems* to contribute to supply-chain integration;
3. The *ability to use local research infrastructure* to solve some technology problems which are not addressed by in-house long-term R&D.

There is therefore more of a bias in these enterprises towards issues related to the manufacturing operations dimensions of the broadly defined technology management agenda which was set out in the first section of this chapter. This contrasts with the technology-intensive firms whose technology management problems are more firmly located within the field of optimizing the process of product innovation.

In both types of large enterprise—technology-intensive and technology-contingent—it is clear that some firms in Europe are at the leading edge of world best practice in the various aspects of technology management we have identified (though it is doubtful if many firms are world class in all aspects). But, as might be expected with the diffusion of organizational innovations just as with technological innovations, there is a large group of firms which fall significantly short of best practice on some or all of these new techniques. It is therefore clear that these firms, and the economies to which they contribute, could benefit from changes in their practice.

The Technology Management Deficits Explored

We now examine these deficits in more detail and explore the background issues and the public policy implications. The following three issues are explored:

1. *The need for many large firms to change their organizational practices in order to utilize technology more effectively.* In particular, the need to systematically employ cross-functional management techniques which make the innovation process more effective and efficient, and which make the operations of a company both efficient and flexible.
2. *The need for large firms to have strategic 'maps' of the technologies they intend to utilize in the future*, and how they will access those technologies, and how their maps relate to those of collaborating institutions in the private and public sectors.
3. *The increasing need for firms to access a greater proportion of their technology from outside their own organization rather than to produce it in-house.*

Cross-functional management

This is a short-hand term which we are using here to refer to a wide range of changes which are emerging in the organization of enterprises across all the advanced economies. An important feature of recent research and popular writing on the management of business firms is the identification of a number of new approaches to organization and co-ordination. The distinctive characteristic of this new approach is a movement away from vertical hierarchical modes of organization towards more horizontal and network-based modes of organization (Sproull and Keisler, 1991; Quinn, 1993.) This trend underpins a variety of more specific changes in management practice such as 'de-layering', 'flat structures', and 'empowerment', and is closely linked to more elaborate and wide-ranging ideas such as lean manufacturing. A central assumption of this trend is that decades of excessive functional specialization and detailed division of labour in organizations have restricted the ability of employees to perform to full potential, and have reduced efficiency and effectiveness. The new principle is to create structures which enable employee potential to be 'released' and directed to customer-centred organizational goals.

This trend in the broad context of organization studies resonates with a long-established finding concerning the management practices and techniques necessary to achieve successful technological innovation. Going right back to Burns and Stalker's (1961) concept of the 'organic organization'; and continuing through the classic innovation studies of the 1970s and 1980s (see Freeman, 1974;

Project Sappho (Rothwell *et al.*, 1974); Coombs, Saviotti, and Walsh, 1987), it has been clear that integration of the efforts of technologists, marketing staff, and operations management staff are a prerequisite for effective innovation. In recent times, however, it has become clear that there are a variety of organizational methods for achieving this cross-functional integration, and that some are much less effective than others. Matrix management techniques have been extensively used and experimented with as companies have tried to establish intimate relationships between their market-facing personnel and their technologists. Clark and Fujimoto (1991), for example, distinguish between several variants of the horizontal dimension of the matrix theme ranging from 'light-weight product managers' through 'heavy-weight programme managers' to 'dedicated teams'. The metaphor of the 'rugby team' mode of organization has been counterposed to the 'relay race' as a way of capturing the essence of the new approach. Smith and Reinertsen (1991) provides a concise account of the practical features of the techniques and their potential benefits.

All of these approaches place a greater premium than traditional techniques on the skill-profiles of team members, on their team-working abilities, and on the human resource management and training strategies which underpin these abilities (Andersen, 1993). Implementing the new approaches is therefore not simply a matter of issuing new instructions, but depends on a broader and deeper process of organizational change. Furthermore, there are potential dangers to be avoided in attempting to speed up innovation, such as a bias towards short-term projects, and a loss of control over the long-term accumulation of skills and knowledge (Crawford, 1992). Finally, in the limit, large organizations are sometimes presented as being in transition towards the ideal of the 'network organization' in which smaller units of the organization form (temporary) units which accomplish specific tasks and transact with other such units in order to fulfil their part in complex supply chains.

At the most simple level, these developments in the organization literature reinforce what the innovation literature has long advocated. This is that there is a prime need to involve technologists, marketers, and production staff jointly in all the stages of innovative projects from research, product concept, development, detail design, manufacture, and sales and marketing. The alternative stereotype in which each specialist 'throws the project over the wall' to the next is recognized as a major pitfall which damages the chances of commercial success.

Whilst this principle may be easy to recognize, it is not such a straightforward matter to put in place structures which achieve the necessary co-ordination. In some cases it is approached by the formation of multi-disciplinary teams dedicated to particular projects.

An example of the use of dedicated multi-disciplinary teams is the case of a multinational chemical firm that recognized that its refrigerants business was terminally threatened by the impending regulatory constraints on chlorofluorocarbons. So they set up a special task-force to produce a new product and a corresponding new process in a time-period much reduced compared to their normal development cycle. They had to run many aspects of the research and the process development in parallel rather than in series, taking calculated risks on technical problems being solved in one domain, whilst proceeding in other domains on the assumption that they would be solved. At the same time, investments in production capacity and the establishment of marketing capability were proceeding before the product development was complete. The company succeeded with the development, and is now trying to transpose some of its learning from this team into other product-development activities.[3]

Another approach to maximizing multi-functional co-ordination is to use committee structures which are carefully constructed to reflect a balance of skills and influence from all the relevant functional areas and types of expertise. These committees then plan and monitor the progress of the development and exploitation of technology in innovative projects. A major UK electrical engineering firm, for example, with substantial automotive and aerospace businesses uses a Technology Council with senior representation from all business units and from different functional areas to review and steer the strategic direction of the longer-term components of the company's R&D portfolio. This body is important not simply for the substantive decisions it takes, but at least as much for the 'political' process it performs in allowing the different stake-holders in

[3] Further case studies of the use of dedicated teams for product development can be found in the useful series of articles reporting the findings of the Harvard 'Manufacturing Vision Group' in *Harvard Business Review*, September/October 1994.

technology in the organization to engage in debate and appreciate each others' perspectives.

At the far extreme of this principle, we could regard the recent vogue for business process re-engineering as the radical attempt to fundamentally reduce the autonomy and inertia of specialist functional groups in an organization, and to replace that mode of thinking with a more 'horizontal' approach based on speedy implementation of a linked series of steps which begins with inputs to the organization and finishes with 'customer satisfaction'.[4] In the context of technological innovation this trend manifests itself mainly as an intensified concern with speeding up the time to market in the product-development cycle.[5] This involves using organizational supports to 'parallel' working as well as employing technical supports such as shared databases, IT groupware products, and concurrent engineering approaches such as those pioneered in the aerospace and automotive sectors.

We are not concerned here with whether all of these trends are equally important, or whether they will ever fully materialize. What does seem clear, however, is that there is a discernible trend towards trying to overcome the penalties in efficiency and effectiveness which have resulted from excessive functional specialization in the organization and management of enterprises in advanced economies. What then is the public policy significance of this phenomenon? The answer to this question is clear in principle. If some organizations, and some regions, perfect and deploy these new organizational techniques before others then they may enjoy competitive advantage as a result. Public policy in a particular nation or region therefore has a responsibility to:

- *research* the evolution of these new techniques and to try to help the enterprises themselves to identify effective approaches;
- *disseminate* knowledge about effective techniques;
- *develop training programmes* which meet skill gaps amongst the personnel of the organizations in a region, in order to help organizations to adopt the new techniques more rapidly.

[4] BPR as a concept rose to prominence in the early 1990s, and is often associated with the work of Hammer (1990). For a review and critique of trends in this field see Coombs and Hull (1995).

[5] See for example Smith and Reinertsen (1991).

None of this can be done as a substitute for the actions of the enterprises themselves. However, there is clearly scope for public policy to lubricate and assist in some aspects of the creation and diffusion of these organizational innovations.

Strategic technology mapping

Flat organizations with de-centralized decision-making are adopted because they are presumed to be very responsive to market demand. This may well be true in the short and medium term. However, such organizational structures can exhibit deficiencies in developing and exploiting long-term strategies, and in particular can be weak at building long-term strengths in particular types of resource such as tacit knowledge, human-embodied skills, and particular areas of technological expertise. Indeed this criticism has been levelled at a number of major US companies by Roussel, Saad, and Erickson (1991). They argue that such companies are 'stuck' in a 'second generation' approach to R&D management in which big corporate labs have been either abolished or brought under the direct control of business units. The price of this change, however, is that there is no proper co-ordination between the business unit 'owners' of the R&D effort on the question of longer-term technology acquisition strategy. In order to counteract this, they propose the adoption of 'third generation' R&D management, which puts in place a layer of strategic management of R&D which is concerned with the evolution of the corporate portfolio of technology assets.

The results of our study reveal that some problems of a similar type have also occurred during the 1980s in the management of R&D in large diversified companies in Europe. The total R&D portfolio of a corporation can often be firmly divided into units dedicated to the service of specific strategic business units. This reduces the visibility of technologies which are relevant to more than one business and risks the under-utilization and under-development of such technologies. This leads to both static and dynamic efficiency losses.[6] The dynamic ones are the ones of most interest here. For if the corporate parent of the business units does not take the responsibility to 'map' the distribution of technical

[6] This issue is discussed in more detail in Coombs and Richards (1993*a*).

expertise across its portfolio, and does not form a view of how that portfolio should evolve, then it is effectively removing such matters from the process of corporate strategy-making.

For example, a UK-based materials company with worldwide operations and major business in catalytic systems, ceramic colours, and materials for the electronics sector exhibited some of these problems in the late 1980s and early 1990s. The corporate technical centre was changed from 70 per cent corporate funding to 70 per cent business unit funding. The businesses drove the agenda of the technical centre more short term, and could not co-operate well on agreeing between them on any R&D projects in areas of common interest. This undermined the ability of the technical centre to develop a technology strategy, and the problem was not really visible to the business unit boards or to the corporate board. It took several years for the problem to be recognized and for the balance to be shifted back in the other direction by establishing mechanisms to identify long-term projects, secure consent for them, but prevent the funds from being affected by short-term business unit problems.

This danger of 'excessive decentralization' of R&D is being recognized by some companies, and there is growing interest in 'technology mapping' and technology forecasting as useful tools to aid strategy-making. The connection with the public policy agenda is very clear. Long-term technology needs of large corporations, once they are identified, can play a role in shaping their long-term R&D programmes. Those programmes have complex and synergistic relationships with the publicly funded science and technology base, both in terms of the research output of that science base and in terms of the training role which is delivered through the science base. There are obvious opportunities to leverage scientific effort and to co-ordinate effort through formal and informal collaborations in these strategic R&D topics which are still reasonably far from market. This is an important element in the thinking behind the Foresight exercises which have begun to diffuse more widely through the policy agendas of advanced economies.

Technology Foresight exercises, despite their variations between countries, have as their major common feature the objective of producing a picture of the state of development of a list of specific technologies at some fixed date in the future. They are therefore concerned with creating a 'best guess' of how quickly and in what direction technology will unfold, based on the informed opinions of the technical and commercial experts whose actual decisions will influence that unfolding.[7] This frequently involves the use of such techniques as scenario-writing and Delphi questionnaire exercises in order to capture the varying judgements of the future which different experts may have. The Foresight exercise may be thought of as having two outcomes: a substantive one and a procedural one. The substantive one is a set of statements about the technical and commercial future which can then inform resource allocation decisions in R&D. The procedural outcome is an intensification of discussion of future developments amongst the technical community, and perhaps some degree of convergence of views, which then influences the actors to be more confident in taking steps along the roads mapped out by the Foresight exercise.[8]

It is clear from the experiences of these Foresight programmes that they have a significant network-building and strengthening role. Technologists from different companies and from publicly funded laboratories are drawn together in the committee structures and through the consultation mechanisms to consider the economic significance of various technical options, set in various alternative future scenarios. These joint activities appear to give enhanced legitimacy to the principle of thinking systematically about long-term processes of technical change. It may also be that they have more subterranean effects on the reinforcements between actual decisions on technical matters between companies with overlapping interests.

The public policy significance of technology mapping is therefore two-fold: *it is a further source of competitive advantage* for the enterprises of a nation or region, and therefore merits public policy support in the form of training or other forms of assistance; and *enterprise level technology mapping* provides an important part of the 'partnership' between the major participants in national and regional R&D which is the rationale for public Technology Foresight programmes.

[7] See Martin (1991) for a discussion of the principles of Foresight.

[8] The author observed at first hand this dual purpose of Foresight in the UK exercise in 1994/95. He served as facilitator and consultant to one of the expert panels (Manufacturing, Production and Business Processes) in the UK Foresight programme.

External sourcing of technology

It is widely accepted that even the largest and most technology-intensive enterprises find it increasingly difficult to cover the whole range of technologies which they need to access. There are many reasons for this, but two stand out. First, given products and services tend to develop in a way which involves the number and variety of constituent technologies in each product increasing. Put another way, the level of technological diversity increases faster than the level of product diversity in the output of the firm. Second, a substantial number of strategic radical innovations are of the 'fusion' variety,[9] in which previously unconnected technologies are brought together to provide novel functionality.

In addition to these 'external' pressures on the ability of enterprises to cover the technological waterfront, there is the 'internal' phenomenon of enterprises tending to adopt a tighter focus on core competences. It is increasingly seen as dangerous for enterprises to aim to be totally self-sufficient in all components of their operations. Instead they are encouraged by current management thinking to clearly differentiate between those skills and resources which are proprietary and crucial to their competitive advantage, and those which are secondary and can be provided more effectively by external suppliers or partners.

All of this results in an increase in the tendency for the enterprise to source more of its technology externally, either by direct purchase of technology and research, by collaborations, by strategic choice of suppliers with good embodied technology in purchased inputs, or by strategic acquisitions of companies.

An example of a strategic alliance which is driven by this type of technology sourcing problem is that between an industrial gases company and a major electronics company. The gases company has a division which makes anaesthesia machines for operating theatres, as part of its health-care business. Its competitive edge comes from the excellence of its gas-handling technology and its proprietary anaesthetic agents, which the machines are fine-tuned to deliver. However, they realized they were losing ground to other competitors on the degree of computer integration of their machines into the broader hospital IT environment. They could not

[9] See Kodama (1995) for a detailed discussion of 'fusion' innovations.

resolve this issue internally, and so teamed up with a leading IT company to provide the instrumentation and IT interfaces to keep their machines competitive on all fronts.

The public policy dimensions of this picture are clear.

1. Some of the institutions in the publicly funded science base are suppliers in the growing technology market place. This raises complex issues of balancing the objectives of these agencies.
2. Collaborative technology-producing programmes have been prominent targets for public policy in recent years. The picture painted above suggests that they will probably become even more important.
3. The international dimensions of the phenomenon of technology outsourcing give a further acceleration to what we now regard as the 'familiar' phenomenon of the globalization of the science and technology production system. These international flows of science and technology raise important issues for nation-states concerning their ability to localize within their national borders some reasonable proportion of that science and technology infrastructure and the wealth-creating potential linked to it.

Policy Implications

This section considers policy options in three areas: education and training; support for co-operative inter-firm learning; and the role of public programmes of technology forecasting and foresight.

Education and Training

The deficiencies in enterprise practice which we have identified (failings in cross-functional management, in strategic analysis of technology, and in technology sourcing) are all topics which can in principle be addressed to some extent through education and training. For the purposes of presentation we will first consider degree-level education programmes, and then move on to post-experience training programmes.

Technology management and degree education

Education prior to work experience is clearly a long-term and indirect means of impacting on enterprise practice. However, it is important in the long run, and can increase the receptivity of a developing cohort of the workforce to new ideas and practices. At first degree level, it is not possible to educate students in detail concerning the subtleties of different modes of organizational working, since their real-life experience is not a well-developed resource against which they can test conceptual models. None the less, the principles behind innovation and the commercial significance of technology can be effectively taught as a component of science and engineering degrees, and as part of management degrees. It might also benefit economics students to be exposed to more of these ideas.

At the level of the Master's degree and the doctorate the possibilities are substantially greater. Many students on such programmes do have work experience to draw on, and are anyway more mature and more aware of business realities. Specialist Master's degrees in Technology Management have emerged in the USA and the UK over the past four to five years, and are beginning to take root in the system. It is too early to say what the effects of their products are on the enterprises who employ them. They are clearly a development which deserves close attention. At the Doctoral level, there is also the beginnings of a change. In the most recent White Paper on Science and Technology in the UK (HMSO, 1994), it was proposed that students doing doctorates in science and engineering subjects should receive some education and training in the 'nature of the commercial R&D environment'. The underlying diagnosis is that students are often beginning their careers as industrial technologists with six or seven years of specialist technical training, but unable to understand a balance-sheet; with no conception of what marketing is; and with no expectation of having to ever engage with these matters.[10] The discussion on how to implement this policy in the UK is not yet concluded, but we can expect some changes in the near future.

The MBA degree is a special case of degree-level education in that it is routinely preceded by substantial work experience, often technical in character, and is expressly targeted at practical business skills. In principle, then, it should be the ideal vehicle for the formation of multi-skilled people who can be technically literate

[10] The situation clearly differs from country to country. In some European countries, advance engineering education has always included very substantial elements of business practice. But this has not normally been the case in the UK.

and commercially literate, and also be aware of the current thinking about cross-functional management. Unfortunately, many MBA students are taught little or nothing about the importance of technological innovation, about how companies actually do innovate, or about the more sophisticated issues of strategic analysis of technology or technology sourcing. Indeed it was this critique of the MBA in the USA which led US Business Schools to introduce separate Masters programmes in Technology Management. Other schools, however, are trying to reform the MBA curriculum to include more material on Technology Management.

In the UK, the White Paper referred to above also proposed that a central initiative be taken to put more material on technological innovation into the courses of UK MBA degrees. Consequently, in 1995 the Economic and Social Research Council funded a number of groups of innovation experts to produce transferable teaching materials and packages on innovation, designed to be slotted into existing courses on strategy, on operations management, on marketing, and so on. These will begin to emerge in 1996 and will then be marketed to UK Business Schools as course development aids.

Technology management and post-experience training

The availability of short courses on Technology Management for company employees is limited but growing.[11] The national modular Master's degree in Technology Management developed by the Jupiter Consortium is relevant here. This is a recently launched national scheme designed to allow people to gain an M.Sc. in Technology Management on a part-time basis through the accumulation of credits gained by attending a number of one-week residential modules. These modules cover a wide range of topics from innovation and R&D management through to operations management, project management, marketing, quality management, and courses in financial and accounting aspects of technology. The distinctive feature of the scheme is that it is not based on one university but on a consortium. Universities offer courses as part of the national programme, and students can take courses from amongst the whole menu offered by all the participating universities. This means that they can tailor the program they take, and make use of geographical convenience. The university in which they take the

[11] This section draws mainly on UK experience.

majority of their modules will then be the one which awards their degree. It is clearly a scheme which assumes that employers will be more willing to release experienced employees only on a short-term basis and not for one-year full-time courses. As yet, however, the scheme is developing only slowly in terms of take-up by students.

In addition to participating in this national scheme, a small number of UK business schools with expertise in this field offer courses of between two and five days duration on various aspects of technology management. From time to time the big consulting companies offer one or two day 'executive seminars' in this field. However, as a whole the market is limited by the continuing lack of awareness and demand at the level of the senior managers in companies who make decisions about training needs of their employees.

Of course, some aspects of the issues we are concerned with may well be addressed in training courses which are not 'badged' as Technology Management. Courses on business process reengineering, for example, may well cover material on cross-functional management. However, there is clearly a need to stimulate both the demand and supply sides of the training market specifically related to technology management.

Policy Options in Education and Training

We have identified some deficiencies and possible market failures in areas of education and training which would improve the technology-management capabilities of large enterprises. The resulting policy options for consideration by the EU might therefore include the following:

- a review of the state of technology-management education and training across member states;
- a series of sponsored conferences or workshops involving suppliers and enterprise-level users of technology management training. These would have the objective of identifying topic needs, delivery vehicles, etc., and helping to form networks and partnerships;
- EU sponsorship of the development of teaching materials, and sponsorship of actual courses (through subsidy of fees) in order to stimulate the market.

Support for Inter-Firm Learning

An important finding of the study is the fact that inter-firm learning is one of the most effective transmission methods for these techniques. Indeed the whole issue of 'the learning firm' is one which has come to the fore in recent years.[12] There are a variety of formal and informal methods through which this takes place. Direct collaboration, especially along supply chains, is one of the most obvious. Labour-markets in skilled personnel is another obvious route. More intangible processes of hearsay and 'demonstration effects' are also important however; as are some of the horizontal associations in which business people meet to exchange views and share experiences.

Large firms are in principle more experienced and capable at organizing their learning activities than are small firms. However, our study found that there are still deficiencies in the practice of large firms. First, learning tends to be linked to specific functions within the firm which are more or less highly developed. Thus, technology-intensive high R&D spenders may be very good at accumulating technology-oriented knowledge from external networks, but relatively weaker in maintaining their manufacturing skills at international best practice levels. For high-volume assembler firms, the opposite may be true. Secondly, for all types of firm, the sheer volume of knowledge being accumulated by different sub-units and individual employees presents an increasingly difficult problem in terms of creating process and systems for processing knowledge within the firm and making sure it is made available at the points where it can be effectively used, rather than trapped at its point of capture. This is an increasing area of attention for Information Technology and Information System managers in very large companies, where size actually hampers effective learning.

Much of the problem of improving learning techniques is in the hands of the enterprises themselves, and cannot be circumvented or short-circuited by public policy interventions. Rather, such interventions should perhaps try to work 'with the grain' of such existing mechanisms, by facilitation and augmentation.

A great deal of inter-firm learning is a by-product of normal commercial relationships and is difficult to shape through policy

[12] See for example Dodgson (1993*a* and 1993*b*).

interventions. Some learning however, is organized through 'clubs' or professional associations, and other such standing organizations. Some of these bodies spring up spontaneously in response to a perceived need amongst some managers and enterprises to meet gaps in their knowledge, or to share experience of common problems. This phenomenon appears to be happening with respect to the broadly defined issue of technology management. The most obvious manifestation of this is the development of the International Technology Management Forum, organized by, and aimed principally at, practitioners rather than academics. Four of these events have now taken place. At the 1995 meeting in Helsinki, the theme was the 'integration of technology strategy and business strategy', which reflects a concern with one of the issues discussed earlier in this chapter, namely the shift from 'second to third generation' R&D.

Another form of organization is the subscription club. An example of such a club was formed in the UK in 1991: the Technology Strategy Forum. In 1995 it had twenty-eight corporate members. It meets four times a year for meetings on specific aspects of technology management, and its members are named senior executives with direct involvement in technology management in their companies. In addition to meetings the TSF sponsors small research projects, and engages in informal information exchanges between members. Benchmarking activities on technology management issues have now begun to take place between member companies.

This type of club is clearly meeting a demand, and is only at the beginning of exploring the possible roles it can play. At the time of writing information is still being collated on similar developments in other EU member countries. Furthermore, some of the issues which TSF members are interested in need to be addressed at an international level rather than a national level.

The policy implication of these observations are that there is a need to review the position with respect to such bodies; consult with a range of enterprises about the desirability of more such organizations being created; and facilitate the formation of such organizations in various direct and indirect ways. These tasks could be undertaken by a variety of national and international policy agencies.

Technology Forecasting and Foresight

Technology Foresight programmes are the subject of considerable attention at the current time. They are perceived by some as having played a significant role in the expansion of Japanese capability to produce and exploit technology in recent years. From observation of the current UK Technology Foresight programme it is clear that they have both a substantive and a procedural role.

The substantive role concerns the actual forecasting procedures and the consequent identification of future technological and market opportunities as perceived by enterprises and technical experts. These 'hard outcomes' of the process are then inputs to the decisions shaping both public and private RTD programmes.

The procedural dimension of Foresight concerns the formation or reinforcement of the contact networks between scientists and technologists in the enterprises of a sector; between these individuals and their professional colleagues in universities and research institutes; and between the scientific and technical experts and the commercial and marketing professionals who also play some part in Foresight. This network formation role is important in terms of informally influencing business decisions, lubricating technical information flows, and generally legitimizing the value of forward-looking visions of how technologies and markets may evolve.

The EU already has substantial experience of forecasting technology and market trends at various levels of aggregation. Indeed it could be said that there was a period in the 1980s when the European Commission played an important role in keeping these principles alive at time when they were less in favour with some national governments and some enterprises. The implications of the recent upsurge in interest in forecasting and foresight is that there is now perhaps more scope, and a more receptive climate, for new initiatives in this field. The key ingredients necessary to make such exercises have a positive benefit on enterprise technology-management practice are that the exercise be collaborative across enterprises and government agencies, and that they be medium term and concrete rather than just long-term and vague. If they are medium term, then they can be expected to feed directly into such fields as standards and regulation, training programmes, and complementary investments in infrastructure. Thus there can be important uncertainty-reducing effects on business decisions consequent on joint participation in forecasting medium-term technology developments. This in turn will help to embed technology planning in the business strategy process. This is the essence of improving the strategic dimension of technology management practice in large enterprises.

The policy implication of the foregoing is that the European Commission should review its activities in the area of collaborative forecasting with enterprises; and should then identify useful ways to expand its activities in this field.

Conclusion

This chapter has argued, on the basis of the study findings, that large enterprises in Europe are not all at the best-practice frontier with respect to cross-functional management, technology strategy formation, and external technology networking and sourcing. The main public policy instruments which can impact directly on this problem are in the areas of education and training for technology management, promotion of horizontal inter-firm learning, and expansion of the programmes in collaborative medium-term technology forecasting. The EC already has substantial involvement in these fields, but the results of the study provide the basis for a review and refinement of the targeting of existing programmes, and for the development of new programmes.

4

Technology-Intensive SMEs: *Policies Supporting the Management of Growing Technological Complexity*

Stefan Kuhlmann and Guido Reger

Introduction

Current trends in technological change—growing complexity and proximity to the science base; increasing interdependence between different areas of technology—pose a significant challenge to technology-intensive small and medium-sized enterprises (SMEs) in Western European countries. This chapter examines these issues by taking a close look at the mechanical engineering sector and the software industry as examples of 'technology-intensive' sectors in which SMEs dominate. It becomes apparent that technology-intensive SMEs can absorb these new developments only to a limited extent. This creates some serious problems for current technology policy measures implemented by European countries as well as by the European Union (EU). Future technology policy concepts have to take into account the changed nature of technological innovation processes: in particular technology-intensive SMEs need more integrated technology policy approaches.

This chapter starts with the identification of important technological challenges. Against this background we discuss the specific consequences for the research and technology management of technology-intensive SMEs. Following this we deal with the technology-management problems of SMEs in the mechanical engineering and the software sector[1] in the European Union. The final

[1] The parts concerning the software sector are to a large extent based on the results of the case study on this branch, which was carried out by Franco Malerba and Salvatore Torrisi (Malerba and Torrisi, 1993).

section shows the need for an integrated technology policy concept for these enterprises and develops four dimensions of operation which need to be accommodated in such a concept.

The Challenge: Growing Intensity and Complexity of Technology

Technology Intensity is Growing and Spreading

Competitive products of high-wage economies are increasingly (high) technology-based. More and more industrial sectors and corporate business areas in the 'Triad' countries are experiencing stronger market pressure to improve the cost–benefit ratio of their products through accelerated technological innovation. This applies not only to research and development (R&D) intensive areas such as the pharmaceutical sector, but also to 'traditional' medium-tech SME-dominated sectors whose technological dynamics have up to now been characterized more by 'D' than by 'R', and which have been more customer- than technology-driven, e.g. the mechanical engineering or the software sector.

Recent statistics show that during the past decade the 'technology-intensive' industrial sectors of the leading industrial countries have tended to increase their 'technology intensity', measured as R&D expenditure in proportion to turnover or production (see European Commission, 1994, 68 ff.). In 1990 technology-intensive industries (defined as those with more than 3.5 per cent R&D expenditure) accounted for 43.3 per cent of the total industrial employment in the European Community (Germany: 54.7; France and UK 41.6; Italy 41.9) compared to 41.8 per cent in the USA and 43.1 per cent in Japan (cf. Gehrke and Grupp, 1994; 70 ff.; see also Grupp, 1995). The relative role of SMEs within technology-intensive industries is statistically still not sufficiently clear; nevertheless there is much indirect evidence to suggest that SMEs are a functionally indispensable element of the technological dynamics of these industries (see e.g. Acs and Audretsch, 1990; Dodgson, 1990; Rothwell and Dodgson, 1992; Wolff *et al.*, 1994). Furthermore, a recent innovation survey in German manufacturing industries underlines the relatively strong innovation orientation of SMEs in many (but not all) technology-intensive sectors, e.g. chemicals, computers, and office machines, electrical machinery, and mechanical engineering (Felder *et al.*, 1994, 42 ff.).

Technology Complexity is Growing and Spreading

Competitive products increasingly have to encompass heterogeneous technological elements ('technology fusion'), as a necessary pre-condition for improved product performance and/or as a consequence of advanced 'systemization' of product clusters. Kodama (1995: 200) defines the technology fusion approach as 'combining existing technologies into hybrid technologies', using mechatronics and fibre-optic communication systems as examples. An OECD report, analysing technology fusion in the field of opto-electronics, states that 'merging different techniques is a major source of innovation' (OECD, 1993a, 7). This holds true in particular for the new generic technologies which are expected to affect products and production processes in many industrial sectors. A futures-orientated study of science and technology carried out in Germany clustered eighty-seven individual technologies around nine generic areas which will be of crucial importance over the next ten years (Grupp, 1994: 388–91; see also Breiner, Cuhls, and Grupp, 1994): advanced materials; nanotechnology; microelectronics; photonics; microsystems technology; software and simulation technologies; molecular electronics; cellular biotechnology; information, production, and management engineering. The study highlights the structural changes which these technologies themselves are undergoing, and it attests to trends of integration between what used to be separate branches of knowledge, as well as to the emergence of new disciplines and their growing interpenetration (Grupp 1994: 379). It was based on assessments made by a large number of technology experts (engineers and technologists). More qualitatively oriented analyses of the dynamics of science and research come to similar conclusions. Schmoch *et al.* (1994*a*) provide case studies in the fields of medical lasers and of neural networks: both areas are characterized by the emergence of a techno-scientific community which establishes an interdisciplinary network of industrial and academic actors (Schmoch *et al.*, 1994*a*: 91). Gibbons *et al.* (1994, 27 ff.) argue that transdiciplinarity will be the privileged form of knowledge production in the near future:[2] 'the transdisciplinary mode

[2] Gibbons *et al.* (1994) argue that a transformation of the mode of knowledge production is going on, a movement from from 'Mode 1' to 'Mode 2'. 'In Mode 1 problems are set and solved in a context governed by the largely academic interests of a specific community. By contrast, Mode 2 knowledge production is carried out

consists in a continuous linking and relinking, in specific clusterings and configurations of knowledge which is brought together on a temporary basis in specific contexts of application' (Gibbons *et al.*, 1994: 29; similarly Mittelstraß, 1994).

Taking the 'science base' of an industrial technology as one important indicator for the technological complexity of the relevant industries, it can be shown that many industrial sectors which used to be considered as 'medium-tech-oriented' are increasingly being affected by science-based technological developments. One can draw a map of technology clusters and of their science base (see Figure 4.1). The map can be read by analogy with a normal geographical map and is based on patent data, manipulated by means of multi-dimensional scaling techniques. The map shows the extent to which patents are officially classified into two or more technological areas (multiple classification). Short distances correspond to high technological similarity, represented by high rates of multiple classification of the respective areas (for the underlying methodology, see Grupp and Schmoch, 1992).

In the upper left quadrant of our technological map one can easily identify the mechanical engineering cluster, while the lower right quadrant shows several smaller 'information technology' clusters (data processing, measurement and control technologies, telecommunications) which are strongly related to the software sector (software does not appear in the map since it cannot be patented). On a time axis we could show how these clusters are moving, how certain elements tend to overlap (merge), and that the density of certain clusters is increasing.

The intensity of the hatching of the technological zones in our map shows the relative significance of their science base (cf. Grupp and Schmoch, 1992; Schmoch, Grupp, and Laube, 1994*b*):[3] the

in a context of application. Mode 1 is disciplinary while Mode 2 is transdisciplinary. Mode 1 is characterised by homogeneity, Mode 2 by heterogeneity. Organisationally, Mode 1 is hierarchical and tends to preserve its form, while Mode 2 is more heterarchical and transient.' (Gibbons *et al.*, 1994, 3).

[3] The science base of technologies is measured by the number of non-patent literature (NPL) citations in patent documents. See for details Grupp and Schmoch 1992, 78 ff., and Schmoch *et al.* 1993*a*. It can be shown that the NPL indicator, i.e. the 'scientification' of patented technologies, was growing during the 1980s worldwide (Grupp and Schmoch 1992, 120), which is a result of growing shares of science-based technologies. Because it is impossible to measure objectively a 'high' or a 'low' science base, the world average of NPL citations was taken as a distinctive mark: technology clusters with values above average were defined as highly science based and those with below average scores as less science-based.

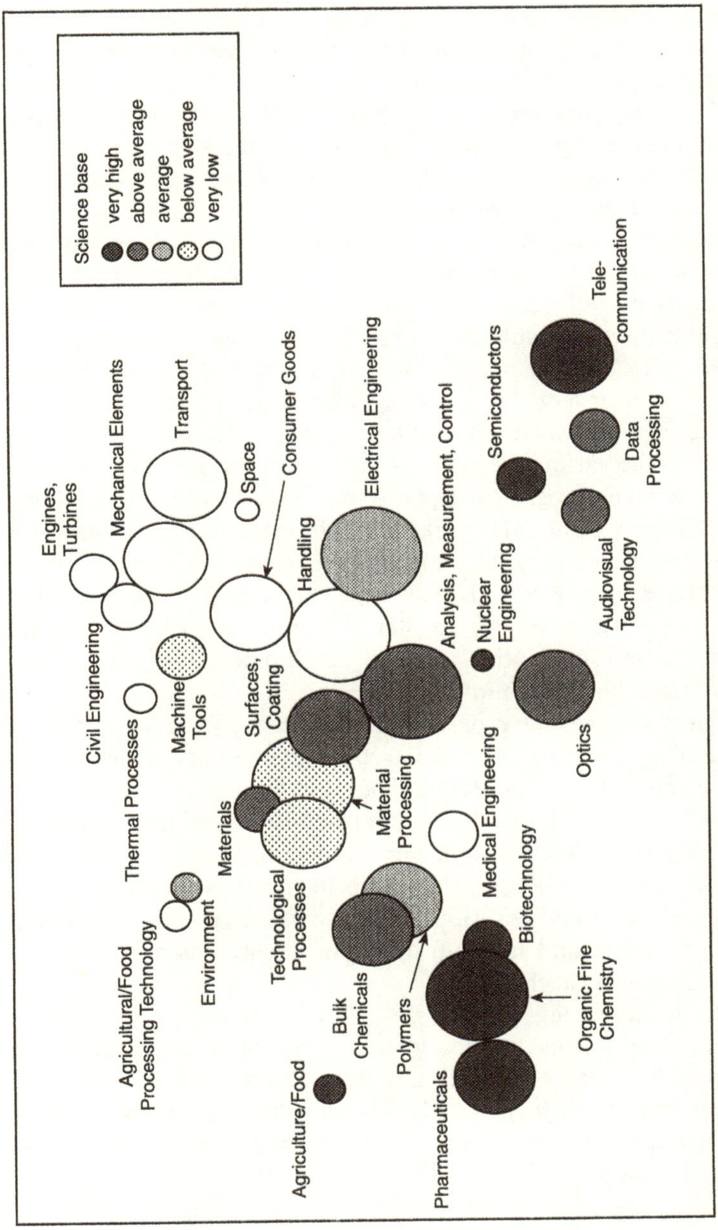

Fig. 4.1. A map of technology clusters and their science base

mechanical engineering cluster traditionally had a below average science base, but it can be seen that new materials (above average science base) are beginning to play an important role. Furthermore, we know from case studies in this area (Reger and Kungl, 1993) that linkages to the measurement and control, the data-processing, and the telecommunications clusters (all above average science base) are coming increasingly into play. Similar developments, but at a higher level of science base, can be shown to be occurring in the case of the software sector.

These new insights in the structure of technology, spill-over effects, and science dependency have important implications for innovation and technology management. The statistically and empirically identifiable trends indicate what we characterize as the growing technology intensity and complexity of industrial innovations. A linear model of the innovation process is no longer appropriate (if it ever was): innovation proceeds with multiple feedbacks between the various elements of the innovation system (Dosi, 1988; for new management concepts see e.g. Van de Ven, 1988; Roberts, 1988; Reger, Cuhls, and Nick, 1994). In many industrial innovation processes technology fusion means the merging of well-known and mature technologies with elements of new or 'science-based' technologies, organized as an iterative process between scientific research and application-oriented development, drawing substantially from knowledge in both spheres.

These changes have particular significance for SMEs, both the technology-intensive and the technology-contingent SMEs, since these enterprises are suffering from well-known constraints (shortage of risk capital, limited human resources). Technology-intensive SMEs have to tackle several complex and risky technological developments simultaneously; they have to match heterogeneous technologies ('technology fusion', interdisciplinarity); and they have to become aware and take advantage of science-based inputs to the development of technology.

In the following sections we will discuss in more detail what kinds of problems technology-intensive SMEs are facing in this respect. First, we will sketch out the problems from a general technology-management point of view and then we will illustrate our argument using the mechanical engineering and the software sectors as examples.

Technology Management in Technology-Intensive SMEs

Models and Requirements of Technology Management

The technological challenges described above are closely intertwined with changes in the competitive environment. Markets are more and more dominated by demand rather than by supply. Flexible specialization, quick response to consumer demands, and rapid changes in production and assembly facilities as well as in research and development (R&D) are necessary, whereas the life-cycles of products and whole industries are increasingly shortening (Reger, Cuhls, and Nick, 1994). What specific requirements for research and technology management in technology-intensive SMEs result from these technological and economic challenges?

Before discussing detailed problems of technology management within technology-intensive SMEs, it is useful to see whether models exist which can serve as guidelines and stimulus for future technology management and successful innovation. Rothwell (1993: 31) points out that successful innovation and R&D are 'people centred': formal management techniques can enhance the performance of competent managers, but they are no substitute for management of high quality and ability, i.e. innovation is essentially a 'people process' and simply attempting to substitute formal management techniques for managerial talent and entrepreneurial flair is not a viable option. Besides that, innovative success is 'multi-factored': there are no single factor explanations. The competence of balancing and co-ordinating all functions, rather than doing one or two things brilliantly well, will support a successful innovation process.

Rothwell (1991, 1993) presents five generations of perceptions of the innovation process, based on an evaluation of empirical studies and theoretical literature. Each perception is a more or less accurate representation of the 'dominantly perceived model' of innovation during a part of the period extending from the end of the 1950s until today. These models show an evolution from the simple linear sequential technology push ('first generation') and need-pull ideas ('second generation') to more interactive, coupling ones which dominated up to the mid 1980s: in the coupling model ('third generation'), the emphasis of management is on the integration across the R&D–marketing interface. The fourth generation—the 'integ-

marked a shift from considering innovation as a sequential process towards thinking of innovation as consisting of numerous parallel processes and feedback and feed-forward loops. The 'systems integration and networking model' ('fifth generation') can be regarded as a somewhat idealized further development of the fourth generation model. In this model, innovation is seen as a multi-institutional networking process: its most significant feature is the 'electronification' of the innovation process. However, according to the author this model is for most companies still in its evolutionary phase and describes the (near) future of the management of research and technology.

Starting from this model of a successful innovation process, we designed a normative model for the requirements associated with research and technology management in technology-intensive SMEs (see Figure 4.2). With regard to technological competences, an SME has to be able to identify and prioritize the technologies relevant for its business areas and to bring its own mature technologies together with new developments. 'Electronification' of the innovation process means the use of interfunctional linked CAD–CAM systems and integrated supplier-assembler CAD systems, the application of simulation modelling replacing physical prototyping, and the use of expert systems as design aid. Parallel development, working simultaneously across business functions and the integration of R&D and manufacturing technologically by CAD/CAM and organizationally by integrated development teams (so-called 'rugby teams') will reduce the concept-to-launch period. Concepts of modularity are essential for simplifying and speeding up the design of new products and for cost reduction.

The necessary entrepreneurial competences consist of the ability to generate and implement a technology strategy, to integrate this strategy with the business strategy, and to focus on quality and other non-price factors. The need to have a corporate strategy based on an awareness of where the leading edge of technology lies is greater than ever before.

A sophisticated and institutionalized interface management between the firm's functions and different technical disciplines or professions is essential to develop and maintain learning abilities and to support knowledge-generating processes. Externally, the organization learns best through strong linkages to leading-edge customers, intensified and international co-operation in R&D with

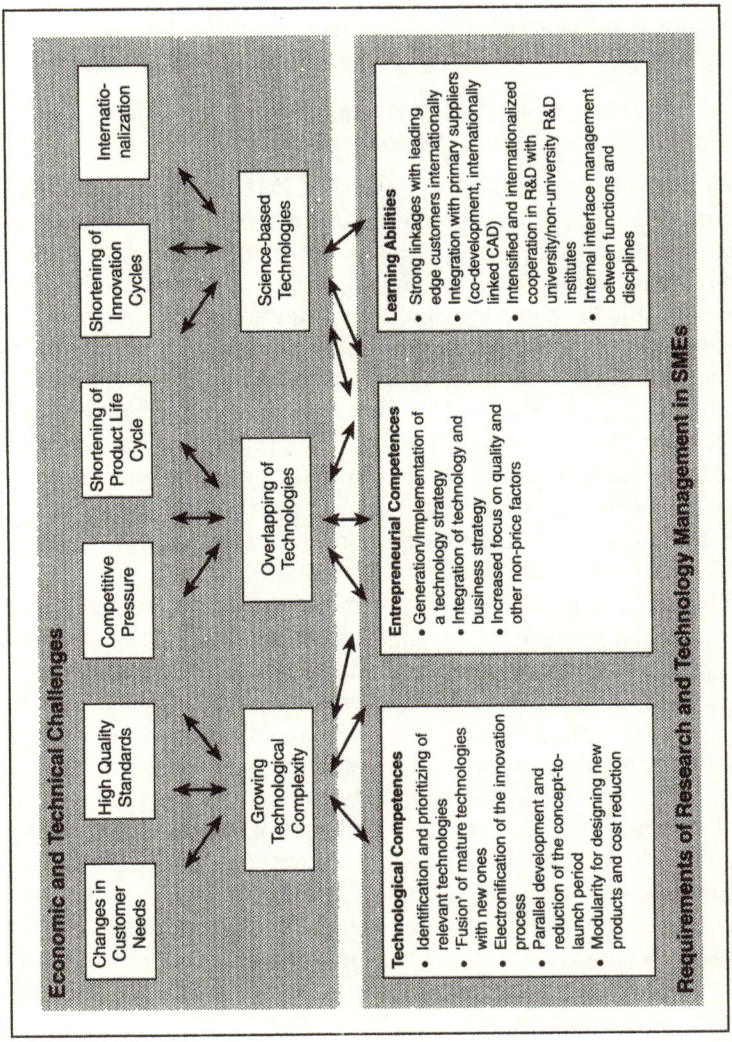

Fig. 4.2. Challenges to and requirements of research and technology management of technology-intensive SMEs

universities or non-university R&D institutions, and through integration with primary suppliers (co-development, linked CAD systems). Since the complexity of the innovation process and the number of actors are increasing, the ability to co-operate and engage in networking are of growing significance for stimulating the knowledge generation and learning ability of the company.

Specialization Capabilities versus the Absorptive Capacity of Technology-Intensive SMEs

Due to their size, technology-intensive SMEs are confronted with characteristic advantages and disadvantages, when tackling the technological and competitive challenges sketched above. Advantages result from the close linkages of SMEs to their customers, which provide learning opportunities and the possibility of meeting specific customer needs. Entrepreneurial and organizational flexibility is higher and can be used as a crucial advantage against large competitors. Disadvantages stem from the inability to cope with the growing technological complexity of new products and the underestimation of opportunities of 'crosspoint' technologies (see Reger & Kungl 1993). As a consequence, strategies for identifying, acquiring, and exploiting crosspoint technologies are not systematically developed and pursued.

Generally speaking, there is a conflict between a company's specialization capability on the one hand and its ability to communicate with an increasingly complex and sophisticated environment on the other. In many cases SMEs are overtaxed to provide the full scope of technological competences required. Technology-intensive SMEs do possess capacity and knowledge to pursue directly new or improved process and product innovations, but this capacity is limited because of their size. It is difficult for them to develop and maintain broader knowledge which allows for the assimilation and exploitation of externally available information. Continuous further training of employees, for instance, is a bottleneck in these firms: this important channel for the acquisition of external knowledge is not effectively followed because the persons to be trained cannot be missed in the day-to-day business.

Cohen and Levinthal (1990) show that the ability to exploit external knowledge is an important component of the firm's innovative capabilities. This ability to evaluate and utilize outside knowledge—they call it 'absorptive capacity'—is largely a function of the level of prior related knowledge. At the most elementary level, this prior knowledge includes, for example, basic skills or a shared language, but it also encompasses knowledge of the most recent scientific or technological trends. Prior related knowledge is the basis on which to recognize the value of new information, assimilate it, and apply it to commercial ends. Following this approach, technology-intensive SMEs may have a structural disadvantage in that they are unable to generate sufficient absorptive capacity for the assimilation and application of new technological trends beyond existing areas.

Taking SMEs of the mechanical engineering sector as an example, there is an obvious difference in the absorptive capacity of companies with an offensive and those with a defensive strategy. Companies pursuing an offensive technology strategy are equipped with a clearly higher degree of absorptive capacity—by definition as well as in practice—as our interviews illustrated (Reger and Kungl, 1993). They invest, in comparison to other SMEs, significantly more in maintaining and expanding their technological competences and their learning ability. In particular, offensive SMEs pay attention to their interface to external technological sources (co-operating companies and public R&D infrastructure). But even so, these efforts are often not efficient in terms of economies of scale and do not easily result in immediately rising sales. Furthermore, the technological paths followed may prove ineffective or less effective than desirable from the perspective of sound investment. For an SME, therefore, pursuing an offensive technology strategy means running a high risk. Public support policies for technologically offensive SMEs, consequently, should aim at lowering such risks, both in financial and in technological terms.

Companies following a more defensive technology strategy, on the other hand, may run a lower risk of investing in the wrong direction, but as a result their absorptive capacity may become inadequate to maintain their competitive position. These companies need to acquire a better understanding of the changing needs of their customers in the future and an increased awareness of the potential performance of new competitors offering more effective and/or less expensive products, using more innovative product concepts and/or production processes. Public support policies for technologically defensive companies, therefore, should aim at stimulating and spreading information on product visions, new markets, innovation strategies, and innovation management. Such policies should facilitate innovation-oriented co-operation with leading companies and the public R&D infrastructure.

Technology-Intensive SMEs in the Mechanical Engineering and Software Sectors

Technological Challenges in the Mechanical Engineering and Software Sectors

Taking a closer look at the mechanical engineering sector and the software sector as examples of 'technology-intensive' sectors (as defined in Chapter 1), growing technology intensity and increasing complexity are crucial challenges to firms in these sectors and can be characterized as follows.

Mechanical engineering

The technological challenges presently facing mechanical engineering firms are mainly represented by the following four areas: microelectronics-based technologies, modular technology, new materials, and the so-called crosspoint technologies. *Microelectronics-based technologies* are the key technologies for mechanical engineering, both in product innovations and in process innovations. The potential applications of microelectronics in manufacturing ('automated factory', 'computer integrated manufacturing (CIM)') and in new products seem to be innumerable and they have a profound effect on the conditions of manufacturing and competition in the sector (see e.g. Fischer, 1990; Hajen, Oppolzer, and Steuer, 1990; Leibinger, 1992; Lay, 1993). These key technologies are crucial for innovative success: microelectronics provides the opportunity to integrate many functions into a single machine and to combine different machines into flexible systems.

The significance of *modular technology* is increasing due to the growth of price-sensitive markets (see e.g. Atkins, 1990; Ehrlenspiel, 1991; Leibinger, 1992; Eversheim *et al.*, 1988). The basic idea is that different machines are constructed by combining a limited number of standardized multifunctional modules in various ways. Modular technology implies that improvements can be made in one module without necessitating changes in the other modules that make up the machine. It can therefore result in accelerating product design and development as well as in cost reductions in manufacturing (because of larger series). Modular technology is an interesting concept for mechanical engineering firms in volume markets which have to realize cost advantages and are oriented towards cost leadership.

The application of *new materials*, mainly constructional materials (e.g. metal alloys, engineering ceramics, engineering plastics, and composites), materials for cutting tools and for coatings, in products and processes is still limited in mechanical engineering compared with other branches, e.g. the automobile industry.[4] This low importance of new materials was also confirmed in the interviews carried out in our study: the lack of internal know-how and the necessity of high investments in new types of machines for forming these new materials are the main impediments to broad application.

The fourth area of technological challenge is the increasing overlap and actual merging of different technological areas such as information and communication technologies, software, optics, laser, microsystems technology, and nanotechnology. The merging of electro-mechanical with software systems seems to be a very challenging step, since software is not only part of the machinery but also reflects a change in the conceptual system (see Brödner, 1993; Atkins, 1990). The share and significance of these 'crosspoint technologies' is growing in mechanical engineering and concerns the firms' own manufacturing processes as well as the technological opportunities for improved and new products. The management of crosspoint technologies is of growing importance for competitiveness, since many mechanical engineering products are already at a mature stage of their technological life-cycle.

The software sector[5]

Apart from the 'software factories' of the largest software producers, many software firms have an *ad hoc* production process. Most firms, especially the small ones, use few structured methods of

[4] See for polymer matrix composites (constructional material) as an example DIW 1989.

[5] The analysis presented in this section is based essentially on Malerba and Torrisi (1993).

production, have relatively unskilled staff, and a loose project management with a weak system of incentives. Important implications of this 'craft' type of production are a relatively low level of productivity (measured in terms of time of project execution), low product quality (measured in terms of the number of errors and the level of maintenance costs), and few opportunities to exploit economies of scale and scope that could arise from the re-use of 'components' (modules of programs) and development tools (methodologies and documentation).

The use of new development techniques is becoming a strategic asset for all software firms. Customization of packaged software and systems integration are the main issues and are activities that result in the development of systems for specific customers. Customization is an activity traditionally carried out by consultants and professional services suppliers. System integration on the contrary is a more complex activity that consists in the integration of different technologies (hardware and standard software packages) into unique systems. This activity also implies the development of custom software and the customization of standard packages.

Customization and systems integration may result in systems which are highly customer-specific. This specificity can easily have a negative effect on other requirements made by the customer, e.g. low costs, improved portability, and quality (few errors and low maintenance costs). Specificity is the most difficult and elusive target for a software firm. Traditionally, software production was based exclusively on the development of programs that were tailored to satisfy a specific customer. With the diffusion of packaged software, customers have given away some degree of specificity to achieve lower product prices. Software firms specialized in packaged software benefit from economies of scale by selling a standard system to a large number of customers (except for some minor modifications). Nevertheless, over the last years the demand for customization of packaged software has also increased. Customers remain interested, however, in the portability of software, a feature that standard packages for personal computers and work-stations have already embodied. They have also become used to the quality of standard packages, which have been improved by solving the questions and problems of thousands of users. Quality for software means a reduced number of errors, and therefore a reduced cost of corrective maintenance of the programs; a high degree of reliability of the programme; and a coherent and friendly system of user interfaces. Modularity is an approach which can help to reply to sometimes conflicting demands for specificity, portability, and quality. Modularity and specificity depend on the size and complexity of the programs. Large, complex programs are usually developed according to a modular approach. System architects design the logical framework of the system and the interconnections among different sub-systems. Different teams of analysts and programmers work in parallel to codify and test different sub-systems that are assembled at the end of the process. Modularity requires the use of structured methodologies and project-management techniques. The possibility and the cost of up-grading, that is of changing the system or a part of it (a module of a program or a sub-system) depends on the techniques utilized to develop the system. The use of new development techniques (such as object-oriented programming languages and tools) allows a higher degree of modification of the system at relatively low costs as compared to the traditional programming languages.

Three consequences can be derived from this trend of increasing user sophistication for software suppliers (see Malerba and Torrisi, 1993): first, producers of packaged software (software houses) are pressed to improve their productivity dramatically in order to survive the fierce price competition. This stimulates firm re-structuring and process innovation and is expected to result in increasing market concentration, as well as diversification of these software houses into customization and system-integration activities. Second, traditional suppliers of professional services and customized software have to improve their productivity and quality because of the competition coming from packaged software. However, they can maintain relatively high profit margins by offering consultancy and training services. Third, systems integrators may increase their market position.

Problems in Technology Management in the Mechanical Engineering and Software Sectors

How do SMEs in both sectors manage the economic and technical challenges and what problems occur in research and technology management? The following comments are based on interviews with thirty-eight European mechanical engineering SMEs and forty-four software SMEs (with less than 500 employees).

Mechanical Engineering

The economic and technological challenges are enhanced by the course of the life-cycle of many mechanical engineering products: generally speaking, mechanical engineering is in the 'transitional phase' and at the beginning of full maturity according to the definition by Abernathy and Utterback (1978). In this phase, the innovation rate is declining, the potential for radical innovations has nearly been exhausted, and process innovations are gaining more and more significance. In response, technology management has to be more and more orientated towards reducing development, engineering, and production costs (e.g. with the help of modular-based technology), but also towards gaining access to new technological areas, extending co-operation, and networking in R&D.

SMEs in the mechanical engineering sector vary considerably in their capacity to manage these challenges. To address this variety, a distinction was made between SMEs with 'offensive' or 'defensive' strategies. SMEs with an R&D expenditure amounting to less than 3.5 per cent of annual turnover (low R&D intensity) were categorized as having a defensive technology strategy whereas firms with more than a 3.5 per cent share of turnover spent on R&D (high R&D intensity) were—according to our definition—pursuing an offensive technology strategy. Firms with a defensive technology strategy represent the conventional type of mechanical engineering firms, but the second type of company is a new phenomenon: they are young SMEs, whose products are located at the interface between the traditional mechanical engineering technology and electronics and information technology.

SMEs with a Defensive Technology Strategy (*Type 1 Firms*), include most manufacturers of special equipment and stand-alone machinery. Their products are supplied to niche markets which offer few prospects of economic growth (see Box 4.1). Single-piece production in small firms hampers the realization of cost reductions and rationalization of the production process. The advantage of these highly specialized producers is their specialist knowledge of technologies and customers. Disadvantages are limited growth opportunities and a very strong dependence on a few specific customers. The focus is on unchanged products, and the main activities of engineers are in the design of customer-oriented solutions. The present technological position of these companies is mainly

ities of engineers are in the design of customer-oriented solutions. The present technological position of these companies is mainly based on past experience. Due to limited financial and personnel resources for R&D activities, there are few attempts to acquire new technological knowledge in core or future areas. Products of this type of firm are marked by technological maturity. To cope with the technological, economic, and organizational challenges these firms are faced with a number of characteristic operational demands. They need to reduce their dependence on a few customers; to expand into international markets, which can be managed by co-operation with other enterprises in the fields of marketing, sales, and common customer services; to acquire external technological know-how and manage this through co-operation with other firms and by training their own engineers; and to try to shift their product supply away from stand-alone machinery towards the supply of systems, at least if they are competing with large producers in volume markets.

SMEs with an Offensive Technology Strategy (*Type 2 Firms*) also tend to be oriented towards specialized products, internal growth, and regional or national markets, but their products contain more components incorporating new technologies (i.e. they belong to the category of NTBFs: 'New Technology Based Firms'). With the orientation towards specific customers, their business strategy aims, on the one hand, at niche markets (niche focus), on the other hand at developing technologically sophisticated products (technology focus) (see Box 4.2). Apart from the development and design of new or improved products, some applied research is also conducted within these firms, with a focus mainly on incremental as opposed to radical product innovations. New innovative ideas are generated inside as well as outside the enterprise, with customers playing a predominant role. But innovations are not only pulled by demand but pushed by technology as well in fields including not just 'classic' mechanics but also electronics, optics, software, information technology, and measuring or control techniques. Contrary to the SMEs of type 1, universities are an important external source of technologies, indeed many of these firms are spin-offs from technical universities, with a consequent strong link between the founder and an academic institute. The most important operational concern of technology management is to maintain the present technological lead and to transform this into economic success.

Box 4.1. *An SME with a Defensive Technology Strategy (Type 1)*

This British firm (less than twenty employees, no formal R&D) produces highly specialized machinery for the textile industry. The firm was founded about 70 years ago, and has been in the same niche market from the beginning. Its competitive strength lies in the specialized engineering skills of the workforce, which enables them to design one-off machines for niche market customers. Its product range is based on progressive development of the original machine, and had remained essentially unchanged in the three years prior to the interview. Many product ideas have arisen in response to customer demand; in the case of substitution of mechanical controls by electrical ones, however, ideas and development were internal to the firm. Product ideas have also stemmed from servicing the machinery of competitors, which provides access to some technological developments. The technological level of present equipment is rather low, with no application of numerical control machinery. Process technology is being improved mainly through upgrading the skills of the workforce, by training and hiring skilled employees. The use of electrical controls resulted in a considerable increase of the skills of the labour-force. There was no formal R&D in 1989 and 1991; R&D is intermittently carried out, with resources organized around products, and the owner-manager is responsible for all technology-related activities. The main obstacles to innovation are lack of financial resources and lack of technical standards.

Although the company has been profitable in recent years, business is very cyclical and highly dependent on the long-term developments of textile manufacturing industry, the more so as the firm has been mainly orientated to the national market, which is in decline. Introduction of synthetic fibres in the textile industry has weakened demand for the firm's products, which are related to natural fibres. As markets have become smaller, the firm has recognized the need to expand into international markets, but no export orders have been obtained so far. Expansion into international markets may be difficult as there is no experience in marketing, which is based on word-of-mouth recommendations. Growth has been internal for many years, but activities have recently been extended by taking over the servicing agency for an Italian firm which produces machinery in the same market segment.

This example shows that a small firm which has a great deal of experience supplying highly specialized markets can be successful (in terms of profit) without sophisticated forms of technology management and without permanently conducting R&D. However the possibilities for growth are very limited, and the firm is highly dependent on developments in the textile industry to which it supplies machinery. Changes in technology and/or demand can easily destroy the economic basis of the enterprise.

Source: Reger and Kungl, 1993.

increasingly with foreign partners to get international access to new technologies. Second, considerable efforts at the level of implementation and servicing are necessary, involving permanent co-ordination with customers to generate and test new ideas during the phase of product development. Third, R&D activities need to be extended more and more to interdisciplinary technological areas—the management of crosspoint technologies is a special challenge for these firms. Fourth, the management of external information and the use of data-banks has to be improved. Finally, distribution channels often need to be improved; a factor which is often neglected because the know-how of many NTBFs is almost entirely of a technical nature.

The software sector

Any future configuration of the software industry will certainly provide room for a large number of firms compared to other sectors. However, there is a minimum ability in terms of efficiency and flexibility that European firms have to gain in order to survive in the market. An important step in reaching this critical threshold of efficiency and flexibility will be the use of software engineering techniques that allow a high degree of re-usability of tools, methods, and documentation, and reduce the costs of customization. These techniques increase the efficiency and flexibility of software production, in a similar fashion to industrial flexible automation.

Box 4.2. *An SME with an Offensive Technology Strategy* (*Type II*)

This small R&D-intensive German firm (with less than fifty employees) is a supplier of tool-monitoring systems, which are components of machine tools. The firm was established in the mid-1980s with a knowledge base that was founded on work by the local university laboratory. Its competitive strength lies in highly specialized know-how relating to specific elements of product technology. The product technology being very dynamic, 50 per cent of products were introduced or drastically changed during the three years prior to the interview. Ideas for product innovations come from customers but also from inside the firm. R&D resources are organized around projects, and product development is performed in teams under the responsibility of the project leader; the Head of Development co-ordinates the different projects. Procedures to ensure adequate linkages between R&D and other functions are very similar in cases of internal and external ideas. In the case of external ideas, after discussions with the customer a draft design is elaborated by a team made up of members of the board, the development department, and the marketing department. After formulating performance specifications a prototype is developed. The product is tested in the customers' manufacturing process. In the case of internal product ideas customers are involved in the same way in this procedure.

Apart from customers, public laboratories and universities are the most important sources for external ideas and R&D inputs, and for this reason the firm was sited close to such institutions. Relevant product technologies comprise mechanical engineering, software, and electronics. In this niche market, there is strong competition from other German and European manufactures, thus the firm is following a strategy of concentration on one product range, whereby it tries to extend the range of application of the product. In order to increase market share, international expansion is a significant element of the business strategy. Project-oriented R&D co-operation with foreign partners is an important instrument to enter international markets. Methods of sourcing technology outside the enterprise include joint R&D projects, publications and open meetings, patent disclosures and training, and hiring skilled employees. 50 per cent of R&D expenditure is related to process technology. The firm's equipment includes automatic handling systems and automatic diagnosis systems.

The product strategy of this firm, a producer of components, not complete systems, has proved to be successful. In this market, product technology must be very advanced in order to convince machine-tool producers to buy and not to make themselves. The main obstacles to innovation are the shortage of very highly qualified workers and financial resources, and the lack of distribution channels, which is a typical feature of small technology-based firms.

Source: Reger and Kungl, 1993.

However, the introduction of flexible systems of production requires an appropriate relevant organizational adaptation of software firms.

The firms investigated in the software sector (see Malerba and Torrisi, 1993) put a great deal of emphasis on multidisciplinary development teams and mobility of personnel. This shows the importance of intense internal and external interactions. Concerning external relations, demands and ideas coming from the customers are the most important source of innovation (see Box 4.3).

Within the firm, project leaders play a key role in the process of co-ordinating different skills. As expected, R&D and other related activities are decentralized in small firms and are subject to informal communication mechanisms. The most important external channel of acquisition of technology is training and hiring skilled employees. This is consistent with the labour-intensive character of this industry; human-embodied knowledge is the major source of technological learning for software firms. The most critical skills for European software firms are in systems engineering, computer science, ability to interact with users, services, and project management. Skills in process engineering and quality assurance are also very important to manage technological change. Commercialization capabilities, and experience with applications and services, are major bottlenecks for many SMEs. System engineer shortages are

Box 4.3. *A Firm in the Software Sector*

This SME produces software for financial services. The company began to develop customized software for leasing, which was later turned into a package, and then began to develop new packages for other financial services. Its main activity became the sale of these packages as well as the maintenance activity and the supply of new versions.

As it was the first firm to develop software orientated towards financial services in its country, it experienced a boom in sales and grew rapidly. Since market niches in software packages are growing slowly, the firm is now expanding into foreign markets through the creation of new firms and by increasing the range of services offered. It is also diversifying into customized financial services for banks and the development of production control systems. Entry into these new areas is being done in co-operation with other firms.

In the leasing area there is no competition on the domestic market. At the moment, the firm is known to be 'the software house' in this area and is often recommended by its existing clients to new customers. Since its software is regularly updated, there is a close and continuous link with its clients. The management is now trying to adopt a similar strategy (i.e. to occupy the new market before the competitors come in and then lock customers into the existing product base) in other growing niches of financial services. Moreover, in these markets, the firm has the advantage of dealing with its previous clients and of having recent products which incorporate the latest technologies.

The use of advanced systems, methodologies, and tools (innovation in process technology) is an essential component of the firm's competitive strategy. Its main strengths are the quality and reliability of its products, the quality of the services offered, and knowledge of the relevant applications: it knows how to solve customers' management problems and to provide consultancy as well as software solutions. The strategy of 'being first into a new market' seems to work well, as the experiences with the leasing market has shown. Ideas coming from clients are of the utmost importance for the innovative products of this firm.

Source: Malerba and Torrisi, 1993.

important, particularly for firms that operate in less advanced regions of Europe and for services firms.

The Need for Integrated Technology Policy Concepts for Technology-Intensive SMEs

As we have noted, technological innovation is an ambitious and risky venture as technologies become more complex and interconnected. Some of the players in the economy, in particular technology-intensive SMEs, can absorb these new developments only to a limited extent. Paradoxically, this is the result of the factors which give them their specific competitive advantages, i.e. their high level of technological specialization. In these circumstances it can be argued that external support is needed to strengthen their technology-management capacity.

Integrated Technology Policy Concepts

Although there is little controversy about the contribution of innovation and of technology management to economic performance, the question arises as to what role state technology policy should play in this area. It is generally acknowledged that policy should not call into question the responsibility of companies for decisions concerning innovations and investments (Fritsch, 1995). It is, however, generally regarded as the duty of the state to maintain a public research infrastructure, both inside and outside the universities. If a complete separability of industrial innovations from the public research infrastructure were to exist, a sharp dividing-line could be drawn between public and private spheres of responsibility. However, we know that technology-intensive SMEs in particular interact in multiple ways with partners within their technology cluster—be it companies along the supplier–customer chain or organizations forming the public R&D and technology transfer infrastructure. Communications within these networks determine significantly the technology management strategies and innovation performance of SMEs. Against the background of the phenomena of technological change discussed above—the spreading of 'technology fusion', the importance of the science link, and the transdisciplinarity of innovations—the dividing-line between private and public responsibilities has come increasingly under discussion. In practical terms, innovation and technology policies nowadays go well beyond simply maintaining a public R&D infrastructure.

Rothwell and Dodgson (1992) summarize the evolution of technology policies towards SMEs in the following way. From the early 1970s on, a new kind of policy began to emerge (initially in the USA): innovation policy. It included grants for innovation, the involvement of collective research institutes in product development for individual companies (as opposed to generic problem-solving), and the use of innovation-stimulating public procurement. In the early 1980s technology policy arose: 'it involved the initiation of major national technology programmes in information technology (and to a lesser extent biotechnology)', focusing on firm–firm or firm–public research institutes collaborative pre-competitive research. 'A further major policy shift has been the change in emphasis from large to small firms . . . By the early 1980s many instruments were in place to support innovation by SMEs' (Rothwell and Dodgson, 1992: 228).

An OECD report on the technological competitiveness of SMEs differentiates two levels of policy action stimulating innovation by small and medium-sized companies (OECD, 1993*b*). The first is the level of the firm i.e. measures helping SMEs to implement development and competition strategies, especially steps aiming to improve their information research capability in relation to technology, their ability to evaluate and apply this information using the appropriate new technology, and their capacity to acquire supplementary resources. The second is the level of the firm's environment, i.e. measures for establishing systematic links between technological research, development, and applications by means of appropriate training and information (the role of innovative milieux, networks, etc.), and programmes for improving the supply to SMEs of the supplementary resources (provision of start-up or venture capital, technology-management advisory services, training facilities, etc.) which will enable them to acquire and make use of technology.

The OECD report shows that a very broad variety of measures have been taken in different countries, and by policy agents at national and regional levels. National programmes include (1) measures dealing with the actual dissemination of technology to SMEs, (2) programmes covering the transfer of information, and (3) programmes aimed at the recipients of information within the

SMEs (OECD, 1993*b*; 73–5). On an international scale the European Union began to implement its own genuinely SME-oriented programmes in the late 1980s (e.g. SPRINT, CRAFT, MINT).

We argue that future technology policy concepts have to go beyond this well-developed and proven range of technology policy measures to take into account the changing nature of technological innovation processes sketched above. In particular, technology-intensive SMEs need technology policy approaches better tailored to their requirements. In developing such policy concepts four dimensions need to be addressed which are depicted in Figure 4.3: (1) visions of applications, (2) new markets and socio-economic needs, (3) future technologies and (4) modern R&D systems (cf. Schmoch *et al.*, 1993*a*). The approach could be called 'integrated technology policy'.

First dimension: visions of applications

This involves the elaboration of visible aims of the future techno-economic development which can serve as guide-posts commonly accepted by society, industry, and the scientific community. One can cite the vision of the 'clean car' (or more generally, a clean environment) as an example, which according to recent science and technology 'foresight studies' (e.g. Breiner, Cuhls, and Grupp, 1994; Cuhls and Kuwahara, 1994) has achieved a high level of consensus in all groups of society, including industrial researchers. Such visions can be elaborated through instruments such as technology assessment and technology forecasting. An important task for the public authorities is the management and moderation of strategic public dialogues and policy formation processes, with the active participation of technology-intensive SMEs. We know from our case studies and from the evaluation of SME-focused technology policy measures (e.g. Kuhlmann and Kuntze, 1991) that many owner-managers of technology-intensive SMEs are very open to new, convincing ideas, and that they often complain about a lack of clear orientation of public policies, e.g. in terms of environmental concerns.

Second dimension: marketable products or processes

The results of public dialogues on societal, environmental, or technological options need to be transformed into marketable products or processes. The market success of innovations, however, depends

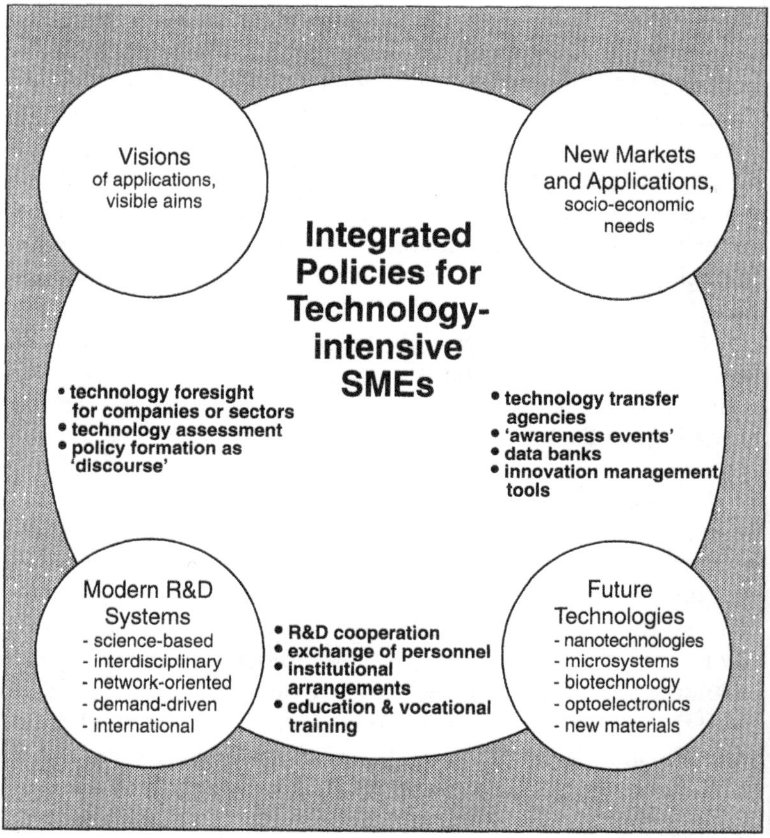

Fig. 4.3. Elements of integrated policies for technology-intensive SMEs

on the existence of an actual demand for these products or processes. The technology management of technology-intensive SMEs needs to include interfaces with such advanced discussions of future needs, and this is an important consideration for policy. Existing technology transfer and dissemination policies, and the mechanisms they adopt (such as transfer units and data banks) are based on a science/technology-push approach, and they need to be supplemented with new instruments which actively stimulate a demand for technological solutions which respond to expressed socio-economic needs.

The Norwegian BUNT programme (Business Development

Using New Technologies, 1989–92) provides an remarkable model for policy action in this dimension: BUNT was to stimulate an increase in the demand for new technologies that would fit into the strategic plans of companies, rather than through the old model of finding companies that might adopt particular technologies. The programme included the training of consultants who carried out strategic analyses in individual companies, aided by a tool-kit developed for this purpose, consisting of about fifty technical notes, theme booklets, and analytical models. These analyses formed the basis of action plans for the implementation of prioritized measures within the companies. This programme can be considered as an instrument for the creation of an innovation and technology-management culture in SMEs. BUNT was evaluated positively and served later as a model for several similar programmes in other countries (see Finne *et al.*, 1995, 17 ff.).

Third dimension: future technology areas

Foresight studies have identified certain future technology areas such as nanotechnology, microsystems technology, biotechnology, opto-electronics, photonics, new materials, or advanced information technology. These strategic areas are science-intensive and will benefit from the integration of contributions from different research institutions.

In such newly emerging areas, where well-established national R&D systems do not yet exist, the potential for an active role of the European Union is significant; technologically offensive SMEs would benefit from this in particular. Because of the complexity of these technologies and the interconnections between technical, organizational, and market aspects of their development and utilization, policy approaches need to involve a number of different complementary elements.

The German programme on Microsystem Technology (MST), launched in 1990 by the Federal Ministry for Research and Technology, provides an interesting model for this dimension of policy action. The programme addressed the systemic character of new intelligent products of technology-intensive SMEs. This was not seen as just a matter of simply combining several areas of technology. Microsystem technology means the co-ordinated usage of modern microtechnologies (e.g. thin-film technologies, micromechanics, integrated optics) and systems technologies (planning methods, simulation technologies, systems architecture, etc.) for the development and production of new types of components and subsystems in response to market demands. Therefore, within the MST Programme non-technical activities (e.g. market analysis, organizational requirements) could also be promoted (Becker, 1995: 140). The programme consisted of three different but co-ordinated types of measures: pre-competitive joint research projects, 'indirect-specific' measures, and technology transfer measures. The pre-competitive R&D joint projects were characterized by significant development risks and financial investments and, on the other hand, by results which could be of interest to a large number of technology users. Joint research projects involved collaborations between research institutes and enterprises in order not only to contribute to technological development, but at the same time to accelerate know-how transfer between science and industry (see Eschenbach, Hafkesbrink, and Lütz, 1995). The 'indirect-specific' measure was meant to advance the widespread practical utilization of miniaturization and system technologies in SMEs. The measure is indirect in that it supports all qualifying firms without further selection; it is specific in that it focuses on a very specific technology. The technology transfer initiative could be defined as accompanying (innovation-supporting) activities for technology diffusion in SMEs. The general aims were to inform firms about the results of technological developments, especially their maturity, as well as on experiences in innovation management, and to actively set up co-operations among firms as well as between firms and research institutions. The transfer initiative included the setting-up of special demonstration and service centres, the promotion of congresses, exhibitions, and infrastructure projects (e.g. international relations in microsystems technology), as well as projects in technology assessment (see Hafkesbrink and Horst, 1995).

Fourth dimension: structures of modern public R&D systems

This dimension concerns the structures of modern public R&D systems (university laboratories, national research centres, research and technology organizations), since they are an important partner for technology-intensive SMEs. These systems are fundamentally affected by the growing importance of science-based technologies and interdisciplinary research topics. Their R&D activities have to become more network-oriented and international, and they not only have to address technological potentials, but also have to become increasingly demand-driven and in this way more open also for SMEs with less offensive technology strategies. Kuhlmann and Holland (1995) developed nine success factors and five performance criteria for publicly co-funded R&D institutes collaborating with technology-intensive SMEs in Baden-Württemberg, a German federal state with an extremely high density of public R&D labs. The performance criteria will in the future be applied by the Baden-Württemberg Ministry of Economic Affairs. They are: the coherence of strategic business area planning and the need of the industrial sector for technological support; techno-scientific competence; success in finding economically viable solutions; the earnings base; and human resources and techno-scientific equipment.

The first criterion, for example, will be evaluated by raising questions like the following: are present and foreseeable market developments (suppliers' markets and sellers' markets) taken sufficiently into account by management? Are present and foreseeable technology developments taken sufficiently into account? Does the institute react promptly to (non-technical) global, social, political, and ecological trends by addressing new topics? Does the institute exert an active influence on the technological targets of industry (increasing awareness of technological challenges)? Does the strategic development of business areas (research groups, services offered) meet these challenges? Does the institute's technological and industrial 'networking' (co-operation with SMEs, strategic alliances etc.) meet these challenges?

Outlook

At the European level, besides the already existing instruments of international R&D project co-operation and exchange of personnel, new institutional arrangements could be considered to improve the technology management of technology-intensive SMEs. This does not involve the establishment of new large European research institutes, but rather the building-up of a longer-term institutional network of existing institutions which are competent to provide technological as well as management support to technology-intensive SMEs in selected strategic areas. In contrast to the generally quite short-term orientated R&D programmes, these networks should have a longer-term perspective. They should have the liberty to adjust the aims of their research according to intermediate results, which generally cannot be predicted exactly in areas of strategic research. Furthermore, the establishment of small European institutes with limited own R&D activities and a major task of co-ordination—similar to the model of co-operative research institutes in Germany—should be considered. New institutional arrangements would be an innovation, with all the risks this implies. Nevertheless, it seems to be a promising approach and should be seriously considered (cf. Schmoch *et al.*, 1993*a*: 114). An institutional aspect of growing importance, so far often neglected, concerns the systems of education and vocational training: since people are the most crucial source of innovation, education systems have to be reconsidered and re-shaped as an integrated element of the innovation system. For many fields of innovation the monodisciplinary orientation of today's education systems does not provide the optimal manpower for private enterprise. Technology-intensive SMEs depend to a crucial extent on highly skilled and flexible human resources with emphasis on technology management skills. Taking into account that education is mainly a state task, there are promising opportunities here for effective policy interventions.

To sum up, this leads to new, additional tasks of technology policy. The institutional and thematic structure of the state-supported R&D and technology transfer system has to be critically evaluated. There are several reasons for thinking that even industrially oriented Research and Technology Organizations are no longer adequately prepared to meet the new requirements of advanced innovation processes in both technologically offensive and defensive SMEs. The other crucial task is to interlink the diverse responsibilities of private and public actors more closely, and to trigger a multilateral exchange of ideas which can deal with the increased complexity of the reality of technological change today. In general, we recommend launching predominantly indirect and less direct support mechanisms, since a targeted influencing of the business and technology strategies of SMEs is neither a desirable nor a promising approach. Distortions of competition and free-rider effects cannot be avoided, in particular with technologically defensive SMEs.

5

The Management of Technology in Technology-Contingent SMEs

Ben Dankbaar

For a large number of business enterprises technological change is like the weather: it happens and you cannot do much about it. Such enterprises treat technology as a contingency which appears by surprise and needs to be dealt with if it cannot be avoided. We call such enterprises 'technology-contingent'. In our research, we have operationalized this mainly on the basis of the level of R&D spending by the enterprise.[1] Technology-contingent enterprises can be of all sizes. In the following, however, we concentrate on the small and medium-sized among them. Much of what we will say about the problems facing technology-contingent SMEs will also be valid for larger enterprises. The latter, however, usually have more and other means to deal with these problems.

It is highly questionable whether technology-contingent enterprises will be able to survive in the high-wage environment of Western Europe in the future, especially now that an increasing number of enterprises in low-wage regions can keep up with technological change at a similar rate of adoption and adaptation. Technology-contingent enterprises will therefore have to enter a learning path which will eventually bring many of them out of this category, towards the category of technology-intensive enterprises. Not all of them will be able to make this transition (and it may not even be necessary in all cases), but all technology-contingent enterprises will have to learn to monitor technological developments, to

[1] R&D spending levels should be related to the average of the relevant sector. Even in sectors where the average ratio of R&D to sales is high (which are therefore often called 'high-tech' sectors), there are enterprises which spend way below that average and can therefore be considered to follow a technology-contingent strategy.

react quickly to relevant changes, and to renew and improve their products and processes regularly. This chapter discusses the main learning mechanisms available to these enterprises and the policies that could be used to push them in the right direction.

Characterizing Technology-Contingent SMEs

Technology-contingent enterprises are not necessarily technologically incompetent. The larger ones in particular may possess considerable technological competence, which provides the basis on which they can react to technological changes in their business environment. Small technology-contingent firms are usually lacking in this respect. They obviously have some know-how in a limited area, but they lack the capacity to investigate and assess new developments in that area, let alone developments in other fields of technology that may threaten their competitive position. In technology-using sectors especially, where most technological change comes from the outside (e.g. from suppliers of equipment and installations), the difference in 'absorptive capacity' (Cohen and Levinthal, 1990) between large and small technology-contingent firms may be quite large. Our study of the food-processing sector in the less favoured regions of Europe has provided us with insights into the problems facing such enterprises. Many small firms in this sector appear to be the prototype for this type of enterprise.

The reason that many enterprises have been able to survive for years and even decades following a technology-contingent strategy is that they have been operating on quite a stable technological trajectory. They could survive with a purely reactive attitude, which required minimal investment in technological capacities, because technological change in their sector was slow and/or very predictable. We are talking here of firms with mature products, where changes in technology are mainly process changes (e.g. the introduction of more precise or more flexible machinery, higher levels of automation, or new forms of process control) that require some adaptations, but no active management of technological change. Nowadays, there are few niches left, where the economic and technological environment is stable enough to adopt such a wait-and-see attitude. Demands upon these firms are rapidly increasing as a result of economic, social, as well as technological developments (cf. Trommsdorf, 1990).

Technological change has been speeding up everywhere, if only because more scientists and engineers are working on it than ever before in history. Besides that, breakthroughs in generic technology fields, like micro-electronics, materials, and biotechnology, have numerous applications and affect almost every sector and indeed almost every single product and process in the economy. Because of these new generic technologies, traditional technological trajectories, which had been following a steady and almost unchanging path for many years, have been re-invigorated, taking unexpected turns, bringing new players into the field.

Such new entrants are only one aspect of intensifying competition in many traditionally stable environments. On the one hand, large customers (for food processing, we could think of the large grocery retail chains and the large food-processing multinationals) have become more demanding regarding price, quality, and reliability of their suppliers, asking them to invest in new technologies which are necessary to achieve the required performance. On the other hand, these suppliers are increasingly faced by competition from outside (Western) Europe, especially at the low end of the market. Not to invest in new technologies means a direct confrontation with these new competitors with very little chance of success. The result is an ongoing streamlining of supply chains in Europe, with strong pressure on traditional small suppliers to adopt more innovative attitudes.

Paralysis is a characteristic response of many technology-contingent enterprises to these changes. The responsible management lacks the time and often also the inclination to take a long-term perspective on the firm and its technology. This is especially the case in small family-owned businesses, where the founder-proprietor has not found a suitable successor—or has been unwilling to search for one. Lack of time is, of course, also a way of saying that the priorities lie elsewhere. Nevertheless, this is a real problem. Because of the increasing intensity of competition, profit margins are squeezed and there appears little room to attract additional personnel to free management for strategic thinking. The point that additional personnel, especially additional engineering knowledge, may be able to pay for itself is difficult to prove in the individual case and requires strategic orientation on the part of management which is now lacking.

In relation to this, small technology contingent enterprises often point to the lack of an adequate marketing function as an important bottleneck in their business. In the past, they were able to survive by supplying a handful of traditional customers. The ongoing streamlining of supply chains makes these long-term relations less secure and even if their traditional customers are willing to maintain the relationship, new demands are made which can often only be met by increasing the scale of operations of the enterprise. The link between marketing and technology is therefore more direct than it may seem at first sight. Only by the acquisition of new customers (more volume) can the conditions for the adoption of new technologies be created. Besides that, it should be remembered that these small firms often gain access to new technologies through their customers, who may be willing to finance the procurement of expensive machinery and tools. It can be argued, therefore, that the introduction of new customers and new markets will have a profound impact on most of these firms. The introduction of an active marketing function will introduce new demands upon their efficiency and technological competence. This could be especially true for efforts to sell on foreign markets, where different customs and regulations have given rise to a different demand structure and different product features. From this perspective, the marketing function refers here not so much to the ability to sell things, but to the ability to communicate with (new and potential) customers and to translate customer needs into new and improved products.

Co-operation between enterprises has often been presented as a functional alternative to internal growth. The European Union has been promoting co-operation and especially international co-operation in its R&D programmes. Small technology-contingent enterprises, however, tend to be very suspicious of the notion of co-operation, as they are very anxious to maintain their independence. Moreover, if these firms become interested in co-operation with other enterprises, they usually prefer firms from the same country or from a specific supply chain, and technology would quite often not be the focus of such co-operative activities. Common marketing or purchasing activities could be other and more urgent reasons for co-operation.

The 'Supplier-Dominated' Environment

Technology-contingent enterprises can be found in all sectors of the economy. In fact, as Hughes (1988) has pointed out, if we consider the variation of R&D intensity within sectors, the R&D intensities of the lower quartile of firms in some 'high technology' sectors (e.g. pharmaceuticals) are below the median of various medium technology industries (e.g. chemicals or mechanical engineering). Technology-contingent enterprises are most prominent, however, in the so-called supplier-dominated sectors. The problems and challenges facing such enterprises can therefore be seen most clearly if we take a closer look at this particular environment. The designation 'supplier-dominated' (or 'technology-using') for certain sectors refers to the fact that technological change in these sectors is mainly produced by the suppliers of materials and equipment. Enterprises in such sectors therefore have a natural inclination to take a passive attitude towards technological change.

Supplier-dominated sectors are traditional manufacturing sectors producing consumer goods like the clothing and shoe industries, or furniture and wood-working. Large parts of the metalworking sector, producing parts and components for assembler industries, can also be considered supplier-dominated. Agriculture and the building industry are also supplier-dominated. Finally, technological change in large parts of the service sector also shows the typical characteristics of supplier dominance. It should be noted that the designation 'supplier-dominated' does not exclude the possibility that the sector itself consists mostly of suppliers to other industries. Where this is the case, it can happen that not just the suppliers of equipment and materials but also the customers become an important source of technological change, e.g. by imposing quality standards or the use of communications technologies like Electronic Data Interchange (EDI). Altogether, supplier-dominated sectors comprise an important part of the economy and it is not surprising that diffusion of new technologies to these sectors is expected to have an important impact on the competitive performance of the economy.

Technological change in these sectors is or has for a long time been incremental, if only because most of the enterprises are unable to manage radical change. Every now and then, however, completely new products, new materials, and new processes appear on the market. Such radical change is often associated with the entry of new enterprises. New materials and the use of information technology in production systems are currently most relevant for a wide range of supplier-dominated sectors. The use of new materials may also require the use of new machinery. Suppliers of these technologies, often large enterprises, are sometimes tempted to use their own technologies and set up subsidiaries to produce the new products themselves. Such new entries to the industry obviously have a much more active attitude towards technological change than their traditional technology-contingent competitors. In such cases, the latter come under strong pressure to speed up the adoption of new technologies, but often fail in their efforts to do so, because they lack the knowledge and organizational capabilities to accommodate the necessary changes.

Production processes in the supplier-dominated sectors are often relatively small-scale. Mass production techniques are increasingly promoted, however, by large customers of consumer products. Retail giants in groceries, clothing, and furniture have developed their own ideas and concepts about materials and methods to be used and demand from their suppliers that these are to be used. Sometimes, these retailers also carry out considerable research on products and materials as well as on the related customer needs. Retailers may also bring in their own designers. Design is an important part of product innovation in many supplier-dominated industries, which may also involve the use of new materials. Design itself can also make use of new computer-based technologies (CAD).

Because of the characteristics sketched above, several of the supplier-dominated sectors have what may be called a 'dual technology' structure. On the one hand there are traditional technology-contingent enterprises, using age-old techniques in the production of traditional products; on the other hand there are large as well as small technology intensive enterprises that make use of new materials and advanced production technologies to produce competing products with similar functions as the traditional ones (e.g. furniture made of different materials), or new and quite different products which sometimes define new needs (e.g. the expansion of soft drinks and snacks in the food industry).

Supplier-dominated sectors are often called 'mature' sectors. The notion of maturity has originally been used for products which no longer exhibit growth in sales, have become standardized, use little R&D for further development, and/or whose disappearance from the market seems imminent (cf. OECD 1988*a*). The extension of this notion from individual products to industries suggests that mature industries exhibit the presence of a large number of mature products, or a slowdown in the appearance of new products. Slow growth of output and/or unchanging products are of course good reasons for enterprises to treat technology as a contingency. In a mature environment, R&D cannot have a major impact on competitiveness anyway. Instead, a mature industry often has a 'downstream bias', i.e. most of its value added is generated in the later stages of the total value chain, with production capacity playing an important role or maybe even distribution creating most of the value added. In other cases, original design may take an important role, which is however very vulnerable to imitation.

Maturity means that the basic production technologies are usually well known and can be implemented everywhere. Work in the production processes has become simplified and standardized and qualification levels of the workforce are of average or below average level. Price will then play an important role in competition and many firms in these sectors are facing competition from low-wage countries. In the longer run, European firms will not be able to compete on price alone against firms from these countries. If transportation costs are low and market proximity unimportant, they will have to add other features to their products: high quality, original design, added functionality, variety, flexibility, speed of delivery. Improving their products and processes along these lines, however, involves a new and more active attitude towards technological change. Technical as well as organizational adaptations will be necessary, including various applications of information technology.

Depending on the definition, large parts of the economy can be considered mature in one sense or another. Some estimates have gone as far as saying that about 75 per cent of all industrial activities in Western Europe could be considered mature (OECD 1988*a*).[2] These would comprise activities such as leather, footwear,

[2] Felsenstein and Bar-El (1989) have combined several indicators to measure the technological intensity of sectors. They identify four different profiles. The one that comes closest to the traditional understanding of 'high technology' accounts for only 15 per cent of all employees and 5 per cent of all plants in Israeli manufacturing

iron and steel, timber, textiles, clothing, building materials, foundries, paper and board, printing and publishing, basic chemicals, glass, food products, and non-ferrous metals. Whatever the exact figures, these estimates indicate that the employment and income-generating potential of 'mature' industries cannot be missed in advanced economies and any thought of replacing them by 'youthful' or 'high tech' industries is highly unrealistic.

The technological challenges in the mature or supplier-dominated sectors are to a large extent defined by the desire to produce higher quality, greater variety, at higher speed, with shorter lead times, and at lower costs. The introduction of quality-control systems is often accompanied by the introduction of new equipment, requiring knowledge of information technology and sensorics. Similarly, demands for speed and flexibility result in the introduction of logistics systems and the installation of computerized production planning and control systems. Electronic data interchange, for instance, guarantees direct processing of orders received on-line from important customers.

In our investigation, we have interviewed firms in various supplier-dominated sectors. In addition to the interviews, expert papers were written on the construction industry and on the clothing industry in Ireland. A sectoral study was made of food processing firms in the less favoured regions of Europe. Food processing is a traditional supplier-dominated sector with numerous SMEs operating traditional, well-known processes. Increasingly, these firms are coming under competitive pressure from two directions. On the one hand, changes in the Common Agricultural Policy and in world-wide trade relations will lead to competition from low wage countries at the same time that the wages in the less favoured regions of Europe are expected to rise to average European levels. On the other hand, these firms are experiencing increasing competition from large multinational food processing enterprises. They will only be able to react to this competition by the use of more modern production equipment (producing higher quality and a larger variety of products) and by investment in marketing, distribution channels, and the related infrastructure.

industry. Sectors conforming to the 'low technology' and 'craft' (skilled labour) profiles account for almost 50 per cent of the industrial workforce.

Technology-Contingent SMEs in Other Contexts

As noted above, technology-contingent SMEs can be found in all parts of the economy and not just in supplier-dominated industries. In every sector we can find enterprises which treat technology basically as an outside factor to which they may have to react. What is to be considered a technology-contingent posture will differ between sectors, as these characterizations are always relative to other firms in the same sector. Thus, a small, technology-contingent firm in a science-based sector may spend a larger share of its turnover on R&D than a large, technology-intensive firm in a supplier-dominated sector.

Discussions of technology management in the science-based industries tend to be dominated by the problems of large enterprises with sizeable R&D operations. It is important to recognize that there is also a large number of SMEs operating in these sectors. In chemicals, for instance, differences in the practice of technology management between small and large firms appeared to be insignificant in our investigation. There were clear differences, however, between firms with significant R&D activities (technology-intensive) and firms with relatively little in-house research (technology-contingent). The technology-intensive firms (with R&D greater than 5 per cent of turnover) gave a slightly higher score to product-related objectives (cf. Figure 5.1).

In fact, for the technology-intensive firms, the most important objectives have been 'increasing or maintaining market share', with 'lowering production costs' ranking second. On the other hand, for technology-contingent firms (with R&D below 5 per cent of turnover), process-related objectives have been the most important: 'lowering production costs' ranks highest and 'reducing pollution level' comes second. Although reducing the pollution level was the third most important objective for the whole sample, it was of more importance for technology-contingent firms (ranking second) than for technology-intensive firms (ranking fifth). This seems to indicate, as might have been expected, that the technology-contingent firms are lagging behind in productivity and pollution control.

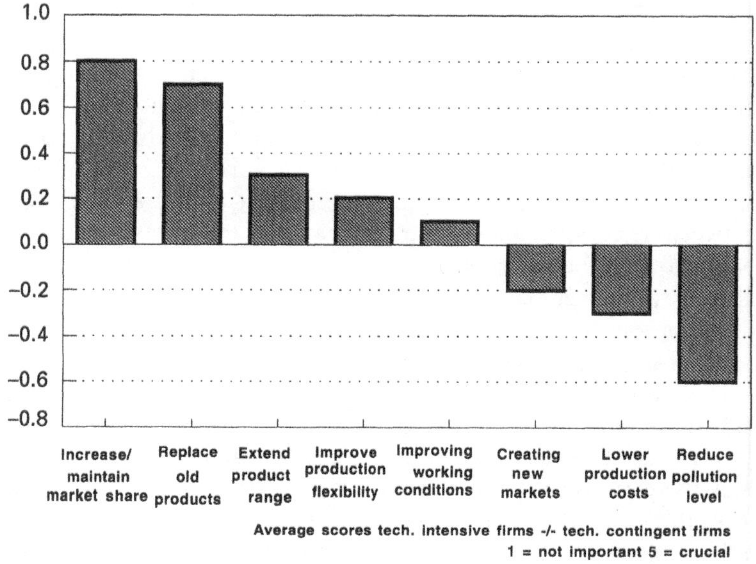

Fig. 5.1. Differences in objectives between technology-intensive and technology-contingent enterprises (Chemicals)

Learning Mechanisms for Technology-Contingent SMEs

How is it that enterprises learn about new technologies? Obviously, R&D activities are one way of learning about new technologies. In the case of technology-contingent SMEs, R&D activities are limited or non-existent. For these enterprises, therefore, the acquisition of technology has to follow different routes. In our investigation we asked all firms (not just the technology-contingent SMEs) to indicate the importance of various methods of acquiring technology from outside the enterprise as opposed to in-house R&D. The following four methods were mentioned most frequently (see also Figure 5.2; the following is based on data from interviews carried out in the Euroregion Maas-Rhine, a region covering Aachen, Limburg, and Wallonia in Germany, the Netherlands, and Belgium respectively):

- *training and/or hiring of skilled employees*
- *joint R&D projects with other enterprises/research laboratories*

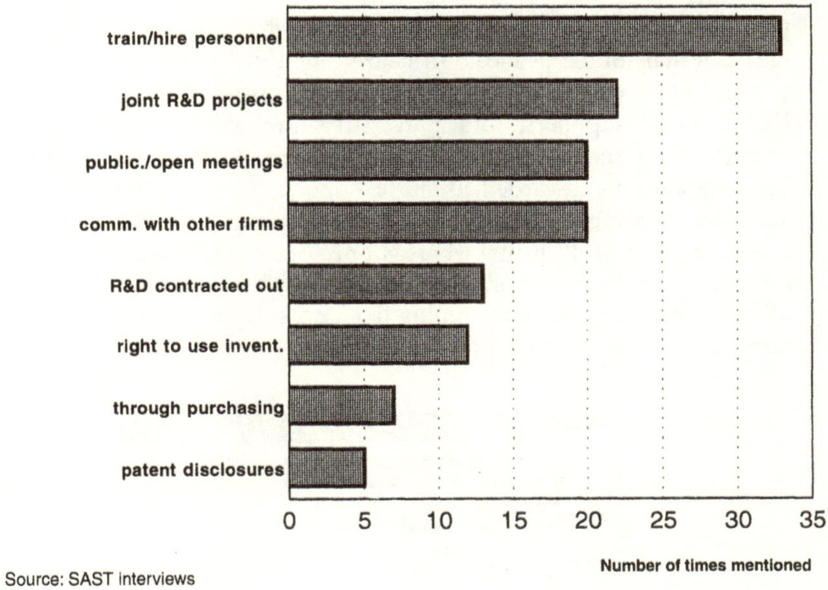

Fig. 5.2. Methods of acquiring technology from outside the firm

- *publications/open meetings*
- *communication with other enterprises.*

Across all enterprises, training and/or hiring of skilled employees is regarded as by far the most important method of acquiring technology from outside. One third of the small companies refer to training and/or hiring of skilled employees as an important method of acquiring technology from outside. This method is far less frequently mentioned by medium-sized and large firms. The chemical firms in the sample also gave less importance to training and/or hiring of skilled employees than one would expect given the importance of a skilled workforce to this industry. The high overall score for training and hiring of skilled employees therefore appears to result mainly from the fact that this is the most frequently used method of knowledge acquisition in technology contingent SMEs, which are of course numerically in the majority.

Firms were also asked to indicate the relative importance of various potential sources of ideas for the actual development or introduction of new or improved products or processes during the past three years. Customers are clearly considered as a very important external source of ideas. In nearly one-third of the cases, customers were mentioned as an important source (Figure 5.3). Competitors, suppliers, and universities are also considered as important external sources for ideas. In general, ideas from outside the company are regarded as more important for the innovative activities than ideas coming from inside the company. This view is shared by half of all firms, whereas only one-quarter indicate that most ideas come from inside the company. The remaining quarter of the firms finds the ideas coming from outside of the firm equally important as ideas coming from the inside.

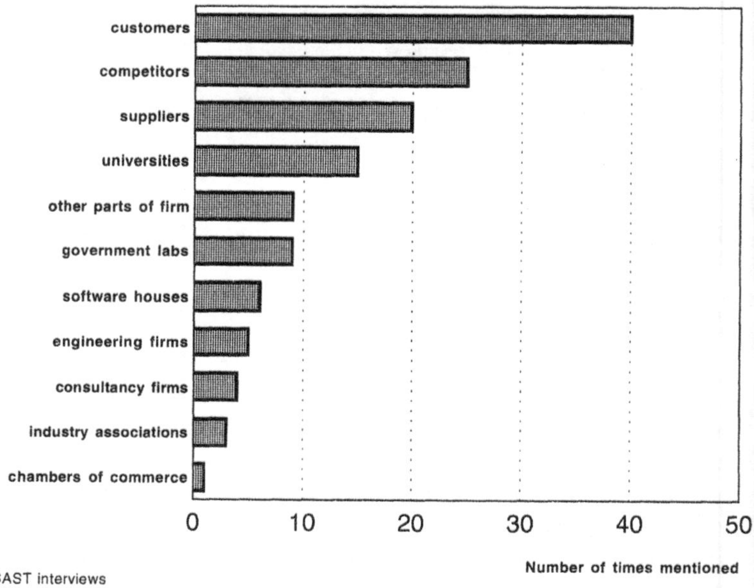

Source: SAST interviews

Fig. 5.3. Sources of ideas for product or process innovation

Small firms consider ideas coming from customers most important. Compared with small enterprises, the large firms find ideas coming from competitors and other parts of the corporation more important. Looking at the size of the firms and their opinions about the importance of ideas coming from inside or outside the firm, not one large company considers ideas coming from inside the corporation as most important. Although this may be a realistic estimation of the relevance of outside ideas, it remains a rather surprising finding, if one considers that most large firms have a substantial research activity that is supposed to generate ideas for the company.

In a representative survey study carried out in the same Euroregion, firms were asked to indicate the importance of various activities for the technological development in their firm.[3] Figure 5.4 shows the results, with separate scores for the different national parts of the region. Patterns do not differ very much between the different subregions, however. Internal technical training is considered most important everywhere. R&D co-operation with third

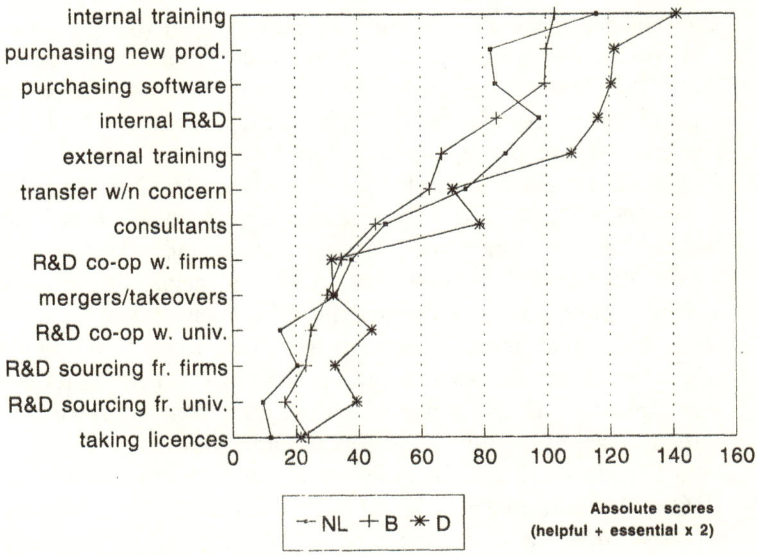

Source: MERIT, ISI, Univ. de Namur

Fig. 5.4. Importance of various activities for technological development in the firm

[3] The questionnaire contained a list of such activities and firms could assess each activity with 'without influence', 'helpful', and 'essential'. For Figure 5.4 a total score per subregion has been constructed by adding the percentage of firms that had indicated 'helpful' to double percentages of firms that indicated 'essential'.

parties is less important for technological change in the enterprises than the use of consultants and knowledge transfer within concerns. The latter are everywhere again considered less important than technical training (in-house or provided by external agencies), in-house R&D, and purchasing of advanced products and software.

The survey study also asked firms to indicate the importance of various business partners for the development of products and processes in the enterprise. In all three parts of the Euregion Maas-Rhine the most important business partners for the development of new products and processes are the customers, followed immediately by the suppliers. Chambers of Commerce, consultancy firms, and university laboratories are considered least important. There are no significant differences between the national parts of the Euregion Maas-Rhine in the ways enterprises value the role of various business partners (Figure 5.5). However, in the German part, the role of universities and other research laboratories are valued somewhat higher than elsewhere, which is due to the special position of the Rhine-Westfalian Technical University (RWTH) in the German part of the region (based in Aachen).

It is interesting to note that, as far as new ideas for products and processes are concerned, competitors are considered more important than engineering firms, software houses, industry associations, or universities. Of course, that does not mean that universities and public laboratories are unimportant for technical change, but it does show that most enterprises get their ideas and knowledge about new technologies and new applications in a roundabout way: not directly from the world of science, but through their contacts with important customers, suppliers, and innovative competitors. Again, this is especially relevant for the technology-contingent SMEs. Without internal capacities for the development or even surveillance of technology, these enterprises depend on their regular business contacts for the acquisition of relevant technologies and innovative ideas.

Recently, the importance of information exchange between enterprises has also been emphasized in studies of dynamic regions like Baden-Württemberg and the so-called 'Third Italy' (Piore and Sabel, 1984; Cooke and Morgan, 1990). Balancing between competition and co-operation, networks of SMEs in these regions can generate the technological dynamism and innovative strength which are usually associated with large enterprises. In our study, we

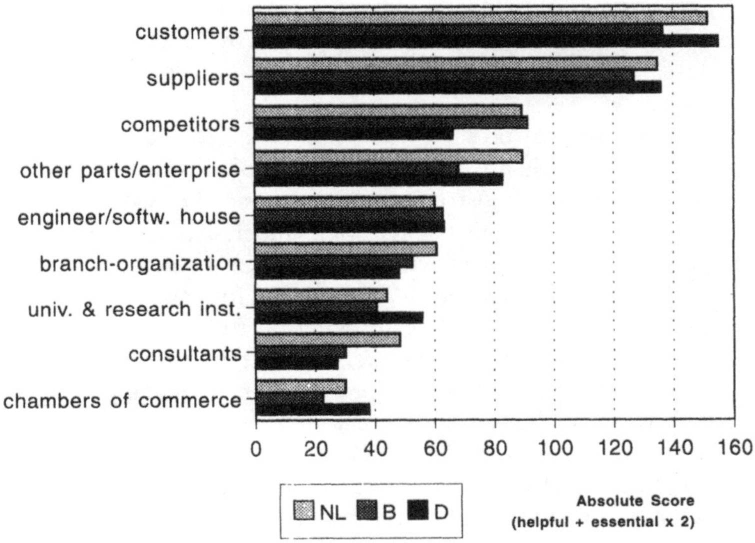

Source: MERIT, ISI, Univ. de Namur

Fig. 5.5. Importance of business partners for the development and implementation of new products or processes

encountered similar effects in the so-called local production communities (LPCs) of Spain. LPCs consist of industrial activities which have developed in a specific geographical environment where co-operative relations play an important role, frequently based on personal contacts. The productive process is divided up over a network of autonomous production centers, each company concentrating on specific production tasks. Decentralization in an industrial branch is thus translated into process specialization of the businesses involved. The decentralized and widespread industrialization observed in Spain has taken place as a result of the presence of these small specialist businesses, integrated in LPCs. Co-operation in the phase of research, manufacturing, and commercialization among different companies emerges as a result of specialization and complementary activities by each firm. Incremental innovations result from progressive adaptation by the companies to the conditions of competition in the market and concentrate on small process and/or product modifications which do not require large investments and therefore are especially appropriate for small businesses. Formal co-operative agreements are not very common. The main formula by which these small businesses access new technological information is through informal exchanges.

Booming growth since 1985 has attracted many multinational companies to Spain. The consequence of this has been increased competition in almost all market sectors. The SMEs in the local production communities have come under strong competitive pressure and are resorting to diversification strategies within the Spanish market. The strategic problem for most local firms now consists in diversifying their product portfolio at very low costs to compete on the Spanish market and at the same time to improve product features, like design, presentation, packaging, etc. in order to have some chances in external markets.

The small firms in the LPCs can be considered as technology-contingent enterprises. However, by operating in a tight network of exchange relations with a large number of other, similar enterprises, these SMEs have been able to improve their absorptive capacity and actually increase their capacity to implement technological change. Whether that is enough to survive in the present situation is another question.

Problems and Bottlenecks for Technology Management in Technology-Contingent SMEs

One could argue that the definition of technology-contingent enterprises excludes the possibility of technology management in such enterprises. If we define technology management as the conscious use of resources to acquire, develop, and apply technology in the enterprise, then it cannot take place in enterprises that treat technological change as a contingency. In fact, we are interested here in enterprises that are moving away from that position. We are concerned with technology-contingent SMEs that are under pressure to take a more active attitude towards technological change. Obviously, that is a process which will take time. Many obstacles and bottlenecks will have to be overcome. In this section we are concerned with identifying such obstacles for the introduction of technology management in enterprises that hitherto have shown little or no evidence of activities in this direction.

Technology management does not necessarily involve the creation of an R&D department. That may or may not happen. Neither does it involve the appointment of a special technology manager. Certainly in small enterprises, it would often be difficult to justify the costs of such a position. It does mean an active support if not active involvement of top management in the systematic monitoring and evaluation of technology fields which are relevant to the enterprise. It does mean a conscious effort to think about the impact of technological change on the market position of the enterprise and a readiness to implement new technologies in the products and processes of the enterprise in order to maintain its competitive position. In other words, it requires of management the taking of a long-term, strategic view of the business.[4]

An important bottleneck for the introduction of technology management in technology-contingent SMEs is the fact that they are often owner-managed (Hilbert and Sperling, 1993). Strategic activities like technology management therefore depend very much on the individual capacities of the owner. These owner-managers often have very limited capabilities to engage in a conscious and active management of technology. Manager-owners often organize their priorities in a different way and then find that they have 'no time' to follow the technological trends impacting on their enterprise. In several cases we could observe that this problem was solved when an older manager-owner was succeeded by a younger person (e.g. his son), who was usually better educated and more versed in the use of information technology.

Technology-contingent enterprises have an incomplete knowledge of available technology. Contrary to widespread opinion, the reason for this lack of knowledge is usually not that this knowledge is not available or is difficult to obtain. Although the supply of relevant knowledge may have been a bottleneck in some regions and some sectors in the past, it is our general impression that this is no longer the case. The real bottleneck now is that many firms lack the absorptive capacity to make use of the various sources of knowledge available to them. Much depends on the ability of the general

[4] The 1994 European Observatory for SMEs points to three key success factors for innovation in SMEs. These are: (1) Enterprises must have a clear commercial strategy and must adapt innovation projects to their strategy, size, and capability; (2) Enterprises must develop clear and precise business plans for innovations; (3) Enterprises, especially the smallest ones, must co-operate, if possible in a transnational partnership (ENSR, 1994: 20).

manager to free himself from part of his duties in order to engage in technological surveillance.

Another obvious solution would be the employment of more technicians and other specialists by these enterprises in order to improve their technological competence. Even if the internal opposition to such newcomers ('we never needed an engineer') can be overcome, it appears that well-educated young engineers and technicians are unwilling to seek employment in such enterprises. There seems to be a widespread shortage of such medium-level technicians all over Europe, most of whom find employment in larger, technically more sophisticated enterprises. Many firms in the supplier-dominated sectors therefore have no other choice but acquiring new knowledge by means of training (internally or externally). This is a time-consuming process, which may last too long for many firms to survive.

One of the main problem for the introduction of more conscious technology management in these enterprises is scale. Scale and the ability to find adequate finance for investments in new equipment are highly related. Many technology-contingent SMEs are simply too small to find the means to follow technological developments, although there are exceptions depending on the precise shape that technology is taking in different sectors.[5] At the same time, larger, more sophisticated enterprises are now using modern, flexible technologies to produce special products and small series that used to come from the SMEs. The need to move towards a larger scale of operations or at least to a larger capital base (e.g. by co-operation and merger) often conflicts with the desire to maintain independence. In the construction sector, there is the additional problem that many small building companies work together on a project basis in constantly changing combinations. With constant uncertainty concerning the practices and demands of future partners, there is little room for the introduction of major planning and communication systems, at least not as long as standard solutions are not available in the sector. The development of such standards, however, requires that some larger actors take the lead (Voordijk, 1994).

[5] In some sectors, e.g. in printing and publishing, modern technology has become available for small-scale as well as large-scale use; in others, e.g. in paper manufacturing, technology cannot be scaled down so easily and therefore has become a prime factor behind growing scale and high financial requirements.

Some of the problems of knowledge acquisition mentioned above are not at all new. Co-operative research institutions and so-called research extension services (technology transfer institutions, e.g. in the agricultural sector) have a long tradition in especially the supplier-dominated sectors, where technology-contingent SMEs predominate. Most of these support services, however, are heavily product-oriented and may not be suitable for the current technological challenges involving information technology. Not much can be gained in this field from co-operation with competitors, who are equally inexperienced. That may be the reason why the role of such traditional technology transfer mechanisms is diminishing compared to that of large suppliers and also customers.

The role of customers as agents for technological change is a relatively new phenomenon.[6] Large customers (often belonging to the so-called scale-intensive sectors, like the automobile industry, the consumer electronics industry, or retailing) are creating networks of supplier firms around them, in which production planning and quality-control systems are implemented. The use of information technology in these relations is imposing new organizational and technological demands on the suppliers. Organized and long-term links with such knowledgeable customers may represent a certain loss of independence for these enterprises, but they gain access to much-needed technology. Being part of a network improves their learning ability.

Policies for Technology Management in Technology-Contingent SMEs

Technology-contingent SMEs will need to develop a longer-term orientation towards their technological environment. The capacity to do so is often not available and the question is, whether or not governments can support the development of such capacity. This is a question which has increasingly received the attention of governments over the past ten or fifteen years. Rothwell and Dodgson (1992) show that governments started to promote innovation in small and medium-sized enterprises from the mid-1970s onwards. In the beginning most attention went to technology-intensive

[6] Von Hippel (1976; 1988) has constantly emphasized the importance of users and customers as sources of innovation.

enterprises and especially the so-called new technology-based firms (NTBFs). More recently, however, attention has shifted to measures promoting technology transfer to what we have called technology-contingent enterprises. These measures are increasingly integrated with regional development policies emphasizing the need for endogenous growth, based on the mobilization and enhancement of regional industrial and technological resources.

Over the past ten years, governments as well as the European Commission have invested heavily in technology-transfer infrastructures. Information on relevant technologies and trends is now widely available to enterprises from such technology-transfer institutions or from other organizations which have become active in this field like branch organizations, chambers of commerce or regional development banks. The main problem now is not availability and accessibility of information but the reluctance of enterprises and their managers to face the need to monitor and search for information on a systematic basis. Part of this reluctance can be related to lack of knowledge and understanding of the relevant trends. This is one of the major dilemmas faced by technology-transfer agencies: the firms that most need their assistance are least likely to come to them.

It is unlikely that such reluctant managers will be convinced by more information offers, computerized databases, or glossy brochures. The willingness to engage in some form of technology management will have to come from a clear reading of market signals. If it is true that technology management will be necessary for most enterprises in Western Europe in order to survive, then it will be important that the relevant signals come to the SMEs as loud, as clear, as frequent, and as soon as possible.

Knowing that something needs to be done is not enough. The enterprises also need to acquire the ability to do something and above all the ability to find out what is to be done. The highest priority for technology-contingent enterprises is probably to increase their in-house technological competence. The more knowledge is available inside the company, the higher the capacity to absorb additional knowledge. The main means to increase absorptive capacity are the employment of technicians and engineers (and giving them time to think about the future of the business), training of the workforce, and last but not least training of management. As noted above, training and hiring of skilled employees remains the single most important mechanism for the acquisition of new knowledge by enterprises in general and by technology-contingent enterprises in particular. Nevertheless, technology-contingent SMEs often find themselves in a position where they feel they cannot afford to employ an expensive technician. Moreover, as long as they do not employ qualified personnel, the need to do so may not be immediately visible.

Governments could undertake action to break this deadlock and encourage SMEs to employ more and better qualified personnel. An interesting example is the 'KIM' programme of the Dutch government. KIM stands for 'Knowledge bearers In small and Medium-sized companies' and the programme supports the employment by SMEs of persons who have just finished higher vocational training.

Employing technicians by itself is not technology management. It is creating an important pre-condition for technological learning and technology management. As we have seen, the major vehicle for technological learning in these enterprises is technology transfer in regular business relations, i.e. by demands from customers, examples set by competitors, and suggestions made by suppliers of materials and equipment. Therefore, an obvious way to improve the technological and innovative capabilities of these enterprises is to intensify these relations and the associated technology transfer. Policies in support of networking try to do just that. Basically, what these policies do is to encourage enterprises to exchange information on all relevant issues within the context of existing business relations.

From a policy perspective, three kinds of networking can be usefully distinguished: regional networking, sectoral networking, and networking along the supply chain. Each may require different instruments and/or different public authorities as the carrier of the policy concerned. Regional networking may be related to regional specialization with a strong technological content, but can also be related to co-operation in training and education. The local production communities in Spain can be seen as examples of such networks. Regional authorities would be the logical carriers of policies in support of such networking. Sectoral networking may also be regional, but will more likely be national or even international. Co-operative and/or public research institutes may play an important role here. In supply-chain networking, the focus is usually on a large firm, which can serve as a supplier of technologies and organizational concepts to its suppliers. This may or may not involve a regional focus. In the automobile industry, for instance, we are currently witnessing a movement towards intensification of relations between main suppliers and the assemblers, where the suppliers are also setting up production facilities in the proximity of the car-assembly plants.

For technology-contingent SMEs, regional networking and supply-chain networking seem to be most in accordance with their business horizon and therefore most relevant. A sector is a much more abstract notion than the immediate environment of an enterprise or its concrete business relations with customers and suppliers. Policies for networking will often appeal to a sense of community that is seldom present in a sector, because the various enterprises in the sector are either competing with each other or have little or nothing to do with each other. Nevertheless, sectoral organizations and research institutes can play an important role in the transfer of sector-specific technologies.

Policies for regional networking could focus on labour-market issues and encourage co-operation between enterprises and training institutions in the region. It is important to ensure the involvement of technology-contingent enterprises in such activities, both because of the need to take account of their needs and because of the learning effect of participation in such actions. If there is a certain measure of specialization present in the region, the model of local production communities suggests that regional authorities encourage the coming of even more specialist enterprises (or the creation of spin-offs of existing enterprises), the establishment of an appropriate supportive infrastructure (for training and research), and other institutions that facilitate communications between enterprises without undermining competitive pressures.

As noted above, policies for supply-chain networking may also have a regional basis, but their focus is on the enterprise (usually an assembler) which is at the end of the supply chain (or, if we take into account distribution, at the heart of the supply chain). Policies in support of supply-chain networking aim to strengthen the learning processes that are taking place between assemblers (or more generally large customers) and their suppliers. They may involve support for the transfer of technology between an assembler and his suppliers, in order to enable the supplier to produce to a higher quality or to develop the capacity for product engineering. They may support small suppliers to co-operate in the development and production of complete components, in order to avoid their becoming second-tier suppliers. In the Euroregion Maas-Rhine, the regional development bank has set up a network of automotive suppliers with the active support of the two car assemblers in the region. The network aims to help the supplier firms in their efforts to acquire the necessary quality certification. The same development bank has also supported the creation of a network of suppliers of an important producer of copying machines in the region. The network aims to support the supplier firms to develop the necessary capabilities to be involved in product innovation at the producer of copying machines.

Outlook

The introduction of conscious technology management in technology-contingent enterprises will produce industrial change in two directions. Some firms will be able to develop considerable technological competence on their own, on which they can base their future competitive performance. They become similar to the dynamic small and medium-sized firms that are predominant in what are called 'specialist supplier' sectors like mechanical engineering. Often, this may involve mergers with similar enterprises or more likely a takeover by a larger and technologically more sophisticated competitor. Other firms will become members of one or more supplier networks that are connected to and organized by large assembler firms and that provide them with the knowledge and means to produce to high quality at high speed and with low costs. The introduction of explicit technology management, therefore, changes not just the behaviour of individual enterprises, but also the character of competition and the structure of industries. Policies in support of technology management in technology-contingent enterprises will therefore need to be closely integrated with broader industrial and competition policies.

It is important to emphasize that it will not be necessary for all technology-contingent enterprises to become technology-intensive in order to survive. Nor is it necessary to assume that every enterprise must become 'high tech' in whatever sense of the word. To survive in a globalizing economy, however, all enterprises will need to be able to monitor their business environment, to deal with changes quickly as they arise, and to renew their products and processes as often as necessary. In that sense, all enterprises, even the technology-contingent ones need to become 'learning enterprises' with a capability for technology management.

6

Technology Management and the Role of the Research and Technology Infrastructure

Lena Tsipouri

Introduction

It has been stressed in the earlier chapters that technology management is a strategic function for industry and that the actual choices enterprises make depend not only on the competences they have already acquired (i.e. competence feeds competence), but also on the outside resources to which they have access. Studying technology management from a policy perspective helps to understand not only what capabilities are involved at the level of firms but also how outside resources can further nourish them, thus providing a basis for policies designed to improve technology management. It has also been emphasized that policies do not directly influence technological change, they do so only indirectly by influencing the behaviour of firms.

It is from this perspective that research and technology infrastructure (RTI) is analysed here, as one component, among others, that can influence the technological decisions of firms. The early period of relatively steady and predictable technological advance has come to an end, and has been replaced with much more pervasive and interconnected changes, which have highly disruptive effects as they are accommodated by industry and society. As a consequence, the role of the RTI changes and more is expected from publicly supported institutions than just the production of new knowledge. In an ideal world, the RTI might be expected to substitute for all needs companies cannot cover themselves: not only offer them solutions to their technical problems, advise them on new products and processes, and support them in the introduction of organizational innovations, but also accompany them in acquiring entrepreneurial competences and learning abilities. Under this scenario the challenge becomes so vast that 'success' depends on the RTI becoming more effective in business terms than the firms themselves. Clearly, in the real world, this is not feasible. The RTI can help in eliminating resource bottlenecks affecting technology management, can help firms build up their technology management capabilities, or increase their speed of learning but can hardly provide them with all they need to accommodate technological change. As a general rule the organizations which classically compose the RTI can be very valuable for improving the technological competence of firms, they can to some extent improve entrepreneurial competences but they can do little to influence learning abilities.

It has already been stated in earlier chapters that we do not subscribe to the view that there is only one best way to manage technology. RTI may or may not, therefore, have a substantial role for improving technology management. The specific role of this chapter is to analyse to what extent and under which circumstances the RTI can be of value. In what follows we build our analysis on four basic ideas:

RTI has in some cases proved highly beneficial for improving the performance of associated firms and overall regional or sectoral competitiveness, whereas in others its role has been limited or of little consequence. Emilia Romagna (Porter, 1990) is a typical case where a satisfactory increase of regional competitiveness and growth rates is attributed to the RTI, whereas the automobile sector is a case of a sector which has grown with inter-firm linkages and very limited RTI support. In the less favoured regions of the EU the proliferation of service providers has not been accompanied by the expected increase of manufacturing growth rates. For policy-makers this can be translated into a need to create RTI only in areas where the market leaves substantial gaps, and at the same time make sure that it works on the basis of at least quasi-market criteria.

Scientific literature on technological change and the emerging studies of the national innovation systems converge at least on one point namely that *'national competitiveness is created and sustained by firms and government policies can only be a help but not be a substitute for it'* (Nelson, 1993). In science policy terms, this means that in its role for technology-management support the RTI has a mission deriving from company needs and not scientific excellence *per se*.

If technological knowledge is available and not used, there is a problem of inadequate management of technology in the enterprise. If this is the case the challenge for the RTI is not only to produce more knowledge, but to help accommodate it within enterprises.

Finally, as stressed in previous chapters, companies learn best from other companies and less from sources outside the market. Support service providers need to take this point very seriously when defining their management approach in order to gain confidence from their clients.

The combination of these four statements, which are the basis of this contribution, clearly suggests that *unconditional creation or sustaining of RTI financed by development budgets is a mistake*. The foundation of our approach is that RTI is only an instrument and not a target *per se*. While there are massive and justified doubts in such a statement from an academic point of view (the need for the promotion of knowledge and educational values), for industrial development policy giving priority to demand stimulation rather than the supply of technology transfer is no longer controversial. In other words universities and research centres should continue their academic tasks, to the extent that educational budgets allow for that, but their additional funding by regional funds, as well as the creation of innovation and technology-transfer mechanisms should be analysed from the perspective of their effectiveness for economic development. This only confirms the interdependence between technology, knowledge, and skills. But while guidelines for academic funding are well established and increasingly based on scientometric indicators, regional policies for RTD are more recent and the lack of tradition makes them more vulnerable. The linear model, though rejected, has left deep marks, which still make it difficult for policy-makers to distinguish effectiveness from quality of research results.

From a systems point of view (David and Rosenbloom, 1990), in trying to separate the attributes of quality of technology from its effectiveness for industrial development, the first task is to identify the general conditions and parameters for funding RTI with industrial rather than academic performance criteria. Although in many cases they coincide, they definitely do not always do so, and in some cases they are in direct competition. The second task is to apply the general findings to distinct innovation systems.

Here we attempt to contribute to the first of the above-mentioned tasks and possibly shed some light on how policy-makers should treat the second task. First we make a distinction between definitions, waves, and types of infrastructure mobilized in the name of industrial competitiveness. Then, success and failure stories will be presented with emphasis on analysis and key features rather than a detailed description. Finally the conclusions attempt to generalize the positive and negative results, and suggest a need for part of the infrastructure in each national innovation system to transform its targets and adapt to industrial needs.

The Supply Side: Definitions, Typologies, and Inadequacies[1]

Research and technology infrastructure is defined here as the sum of institutions dealing in some way with the creation and/or diffusion of productive knowledge in a region. Three issues are relevant to this wide definition of the RTI concept: which are the organizations most often encountered? Which classifications can be suggested that would be helpful for technology policy design? And finally, how relevant is the regional context?

The Typical RTI Institutions: A Static Typology

Many organizations influence the creation of knowledge and its relation to industrial performance. Not only the educational system and vocational training schemes, but all industrial restructuring efforts and incentives are in some way related to the way manufacturing activities adapt to technological change. This corresponds to Nelson's idea that 'the concept of a national innovation system is the set of institutions whose interaction determine the innovative performance of national firms' (Nelson, 1993: 6). The rapidly increasing importance of research, technology, and innovation in relation to policy for industrial competitiveness and ultimately welfare have led to the proliferation of a wide variety of such organizations which are often, but not always, based on state initiatives

[1] This contribution has been enriched by the experience gained from the implementation of the RITTS programme of DG XIII–SPRINT.

and public funding. While historically these institutions were first created in the most advanced OECD member states, they were rapidly imitated by all advanced countries and, at a later stage by less favoured regions[2] and developing countries. The European Commission the World Bank, the UN, and other international organizations have strongly favoured and financed policy interventions for RTI creation.

Two dimensions may be considered relevant for a taxonomy of the RTI (see Table 6.1): its ownership (public, non profit, or profit-oriented) and its role in the innovation process (education and skills, R&D, or technology transfer). The system for education and training can be distinguished from the other dimensions, in the sense that although its contribution may be undisputed, it is orientated more to the long-term and has a much broader function than supporting technological change in business. Market forces alone are insufficient to maintain an adequate education and training system, except in a few specific areas (profit-oriented universities and training centres), and it is widely accepted, even by mainstream economists, that this market failure must be corrected by public

Table 6.1. A static typology of RTI organizations

Role in innovation	Ownership		
	Public	Non-profit	Profit-oriented
Education and skills	HEIs, vocational training	Autonomous educational establishment	Private HEIs, profit-oriented training schools
R&D	HEIs	Foundations, smaller autonomous research units	Research labs in private companies
Technology transfer		Research associations, technology parks, incubators, BICs, TT agencies	Technical consulting, engineering, software companies, management consulting (CIM, QC etc.)

[2] We use the term 'Less Favoured Regions' according to the EU definition: a region whose GDP is below 70% of the Union's average.

intervention (high educational budgets). Core regions are generally well endowed in their educational system, while the less-favoured regions (LFRs) constantly strive to increase the number of local higher education institutions (HEIs) and partly overcome their deficiencies by a massive flow of students abroad.

Research and development is widely covered by universities, dedicated research centres, and private funds, where the latter are convinced that in the long run return on investment is positive. Pure research activities, relevant for the science-based firms, are mainly conducted in universities, research centres, and the basic research laboratories of big, transnational enterprises, while applied research and development addresses more process-focused, technology-using enterprises. The latter demands flexible structures, dedicated to a particular group of clients, and thus autonomous units are more appropriate. For similar reasons general public measures are totally unsuitable for technology-transfer activities. Part of these activities are covered by industrially or publicly initiated research associations, while market-driven activities are covered by private consulting companies. The case of technology-transfer activities is typically better supported in rich markets (in both regional and sectoral terms), while the small technology-contingent firms considered in Chapter 5, as well as the RTI relevant for them, find themselves in a vicious circle of deficiencies of both supply and demand.

The bigger the market, and the greater the demand for RTI services (which tends to be linked with more explicit technology management, in our terminology), the more market-driven, profit-making institutions will be sustainable. Conversely, in small developing markets (regions or sectors) more public intervention is needed. In development terms, this implies that policy-makers need to dedicate higher budgets at an initial stage, but institutions will have to gradually increase the share of their budget covered by their clients.

This classification gives a good though static overview of the various forms the infrastructure usually takes. But the real world is more complicated to handle, taking into account the fact that organizations continuously mutate to adapt to their environment. As a consequence in the EU we now have a very wide variety of organizations comprising the RTI, with overlapping targets and often unclear missions. It is thus necessary to search for additional dimensions, which can shed light on the dynamics of RTI, from the point of view of policy-making.

Typologies of Interest to Policy-Makers

Policy-makers need analytical tools in order to create new RTI structures or evaluate existing ones. Since ownership is exogenous and the role of particular organizations in the innovation process changes over time it is suggested that the organizations involved can be classified according to three additional dimensions. These do not constitute firm categories, but they allow policy-makers to monitor institutions and see to what extent they reflect (*a*) what might be called 'good practice' and (*b*) they cover firms' needs (as further analysed below).

Classification by primary objective and scope

In the past the objectives of RTI organizations were somewhat clearer than they are today, in the sense that HEI and public research centres were dedicated mainly to education (the creation of skills) plus basic and to a lesser extent applied research, while firms' laboratories and research associations were oriented towards development and engineering. The transformation of the economic environment and of the objectives of funding sources, and the abandoning of the linear model in parallel with the recognition of the complexity and chain-link relations of research, development, and engineering, have led to the extension of functions and a wide coverage by each institution independently of its origin. Thus more and more HEIs are expected to get into the technology-transfer business, industrial laboratories to undertake basic research, successful technology-transfer agencies to support small teams engaged in blue sky research to enable them to retain high quality personnel and sustain their state-of-the-art knowledge, etc. In other words the classical objectives blur and the chain-link model can be best read in the variety of functions an RTI organization performs, independently of its origins, primary mission, and status. Depending on their policy orientation, HEIs may now, for instance, fit into any of the cells of our original matrix (Table 6.1), although they predominantly remain in the first. The current picture then shows a transformation of targets and environment, with overlapping missions, unclear boundaries, and often competition among organizations with different original missions within the same national innovation system. The passage from core funding to project funding mechanisms constantly reinforces this competitive climate within the RTI. Thus our first conclusion is *that RTI organizations become competition-driven and their scope may become broader or more specialized as time goes by.* This means that there are technology-management needs which, according to static criteria, are covered by the RTI infrastructure but in practice are not, and vice versa. In our empirical research we often found companies that had their technology transfer needs covered by universities (e.g. in the UK), while technology-transfer agencies had been transformed into training centres (e.g. in Greece).

Classification by size

All of the above-mentioned types of organizations can be found in a wide variety of sizes. While universities tend to be big and multidisciplinary and specialized research institutes or technology-transfer agencies are smaller, their size varies considerably. Conventional wisdom tends to suggest that (1) big, multidisciplinary establishments cannot play an effective role in technology transfer, because they are too rigid and academically oriented, while (2) very small institutions do not offer the critical mass of people necessary to support knowledge production and transfer at a sufficient level of performance.

Evaluation work suggests that the former statement is to a large extent applicable to research centres, which often fail to support their best people, because of the need to spread resources equally among research teams. As a consequence they can become unable to create centres of excellence. The latter statement is applicable to publicly supported, technology-transfer-oriented organizations, such as research organizations or science parks, which do not (for various reasons) reach a size that will allow them to produce synergetic effects. As a consequence, they can degenerate into real-estate measures or publicly supported research laboratories of disputable quality. But although it is possible to give general guidance on how to promote specialization, competence, or excellence instead of multidisciplinarity, as well as on critical mass, *there is no golden rule which determines optimal size* and examples can be found both of success and failure across a broad spectrum of sizes.

Classification by type of client

The older and more widespread types of institutions were created for the provision of public goods, i.e. the education and skills of individuals that will eventually make up the foundations of the national innovation system, or independent research that will add to the national reservoir of knowledge. Research associations, technology parks, etc. on the other hand are normally created to support companies in improving their access to knowledge and adaptation to technological change. Depending on their focus they address specific sectors (e.g. research associations), size of firms (e.g. parks aiming at the attraction of research facilities of multinational corporations), or technologies. Some of these organizations may, by founding law or historical evolution, address a specific segment of firms. Many of them serve exclusively or primarily big companies, because they are the most rewarding clients, both in pecuniary terms and in scientific potential, others have a mixed clientele as far as size is concerned. Some organizations created exclusively for supporting SMEs face serious difficulties in establishing linkages, and proceeding to the stage where they can 'open the door' and subsequently gain the confidence of their clients. Even when they have done so, the results are generally far less spectacular than corresponding collaborations with bigger firms. Empirical studies confirm that the best projects of RTI organizations emerge from collaborations with big companies, leading to science-based innovations, often radical ones. Contacts with SMEs on the other hand lead in the best case to collaborative cost-sharing projects, more often than not trouble shooting and technology-transfer solutions, with little glamour but urgently needed to sustain competitiveness. This applies particularly to the smaller and more technology-contingent SMEs. Thus, *the market pushes RTI organizations towards a preference for undertaking research for large technology-intensive companies, followed by small technology-intensive firms.*

Does the Regional Dimension Matter?

A similar argument applies to the geographical coverage of the RTI. Many organizations are created to serve local or national needs only, while others are more internationally oriented. The best among them can afford to compete in the international arena, work for the most prestigious clients, those more open to technology, and not least the more affluent companies. EU funding opportunities and EUREKA have helped to reinforce this tendency. In other words a kind of 'globalization' of the research infrastructure is occurring, since competent centres all over Europe, independently of their original funding sources and mission, have a tendency to work internationally. In terms of opportunities, specialization possibilities, and prestige, international contracts (and in particular contacts with the USA and Japan) are more rewarding than local ones. Thus technology-intensive firms find it more and more easy to access knowledge outside their own territory, and this supports the view that the regional context has become less relevant. But while international collaborations are adding value, if well designed and balanced with local and national activities, they risk becoming a burden for local development, if they are too strongly oriented towards needs outside their own economy. The best RTI structures in the less favoured regions suffer from this outward orientation, which militates against local economic exploitation of their know-how. In Ireland the RTI, with high quality scientific personnel, has a tendency to work for both the USA and the UK (which involve, of course, no language barriers) more than for the local economy.[3] But the same problem is also applicable to prestigious universities in the UK and even Germany, where contacts with the USA are an indicator of success. This should not, of course, be interpreted as an entirely negative evaluation: centres of competence anywhere in Europe are important assets. Nevertheless, evaluation of their effectiveness must also take account of their geographical coverage, and this sometimes tends to be forgotten.

Unfortunately this globalization of the RTI does not seem to be reciprocated in the behaviour of technology-contingent firms which, lacking the culture to access support-service providers, do not search globally in order to cover their technological needs. Despite public support to create an internationally accessible network of RTI (such as the VALUE Relay Centres established by the EU), they are unable or reluctant to make use of the opportunities on offer. This attitude confirms the view that technology transfer is a social process: the regional dimension matters for creating a cul-

[3] This has led the Irish government to impose maximum quotas for international collaborations in the Programmes for Advanced Technology.

ture of contacts between firms and the RTI. The process of creating initial contacts, which remain predominantly national as the Euroregion experience discussed in Chapter 5 illustrates, takes much longer than that of extending them to global linkages, once firms have acquired this culture.

The above classifications indicate that there is an increasing variety, complexity, and organizational overlap in RTI and explain that there are no clear limits or any single best practice, to suggest how the role of the infrastructure can be optimized for technology management. It is common sense (corroborated by empirical results from evaluation studies (Circa, 1993) to say that what really matters for firms is high quality services and respect for certain rules of minimum size and focus, while from the regional point of view, a good mix of local or national and international activities, as well as coverage of all firms and types of services, is necessary. But these dimensions can only really function as parameters which policy-makers can monitor, in order to make sure that the characteristics of the RTI evolve in the right direction. As there are no standard best practices, the appropriateness of the infrastructure, within the above-mentioned boundaries, has to be determined by the demand side, i.e. the actual and latent needs of firms for RTD support.

Anatomy of Firms' Needs: Effective Versus Latent Demand

Europe experienced a rapid construction of RTI in the 1980s. Through the globalization of firms and technology, access to public knowledge is increasing, and technical know-how is also increasingly available in both academic institutions and bigger companies. Nevertheless, European competitiveness is still constrained by a limited capacity to transform knowledge into new products and processes (Soete and Arundel, 1993) and, as stressed in Chapter 1 of this book, to accommodate technological change. Thus, in accordance with our third assumption, there is a lack of technology management which companies (some at least) need to be helped to overcome.

Effective demand for technology transfer is low. Large, technology-intensive companies, in particular those in science-based sectors and to some extent new technology-based firms, make wide use of RTI and are best served by it. However, the vast majority of European companies, namely traditional SMEs and 'scale intensive enterprises' (Pavitt, 1984), make little, if any, use of it. The actual relevance of RTI for European companies is reflected in the responses to two types of questions posed in our study (see Chapter 5, Figures 5.2 and 5.3). In terms of sources of ideas for innovation, universities were rated fourth, being mentioned less than half as often as customers. Universities, government laboratories, and industry associations together were mentioned less often than competitors. At the same time in the methods for acquiring technology from outside 'R&D contracted out' rates fifth, well below open meetings and common projects with other firms.

But latent demand for RTI services is believed to be very high, since the technological competence of European companies needs to be improved and most traditional companies can be considered to suffer from late adaptation to technological change. The reasons for this demand not being expressed are often associated with industrial culture, a certain reluctance of companies to address unknown environments and not least the justified feeling that RTI organizations are more willing to collaborate with natural winners (large, science-based, international companies) than firms with high needs and moderate performance.

From a technology policy point of view, this implies that two tasks should be attributed to the RTI, the first being to cover the expressed needs and the second, the real challenge in fact, to deal with latent demand and get enterprises to take advantage of the know-how at their disposal. For technology to be transformed into competitive advantage it should not be seen in terms of abstract knowledge but as a regular part of business: it has to be taken out of the hands of scientists and engineers and turned into competitive products. But to convince companies that their entrepreneurial competence and learning ability are insufficient and, more than that, that sources outside the firm itself are able to improve it effectively, is an extremely difficult task.

Using the regional taxonomy of firms presented in Chapter 2, it is clear that large, technology-intensive firms are in fact reasonably well served in their needs, because what really matters for them is the quality of the research and technology organizations, which in Europe is satisfactory. As we have seen, firms of this type have no problem accessing technology globally if it is not available locally. Their technology-management problems and opportunities to improve their learning ability are generally internal affairs and are not further considered as a determining parameter for the creation or behaviour of RTI organizations.[4] On the other hand, large, technology-contingent companies need to improve their strategic perception of technology. Since resources are usually not the main problem, the RTI has a role to play mainly in supportive functions, in the improvement of their learning abilities, and in sensitizing them to the importance of adaptation to change.

Thus our analysis of demand is focused on SMEs, both technology-intensive and technology-contingent, with some reference to the specific needs of big technology-contingent firms. Empirical evidence shows that compared to the USA Europe has less start-up companies, less innovative SMEs, and fewer cases of spectacular growth like Apple, Microsoft, or Novell. It is precisely in this direction that RTI has a role to play. In order to better analyse this, it is suggested that the needs of the three types of companies studied can be classified into various functions (see Table 6.2).

On the basis of this table,[5] a demand–supply matrix of RTI services can be plotted for each region. Any regional innovation system can be represented by placing needs along the horizontal axis and each of the existing RTI organizations down the vertical axis. If all rows of the matrix are adequately covered, then the RTI plays an important role for technology management, if not then the RTI has to be reviewed. Though very different results came out of our regional studies, a typical matrix for an average European region could have the form suggested in Tables 6.3 to 6.5, which analyse the key needs for each type of company and the extent to which they are covered by each local organization, i.e. Higher Educational Institutions (HEI), Research Centres (RC), Research Associations (RA), Technology Parks (TP), Business Information Centres (BICs), Consulting Offices, the banking system, and Others (e.g. imported services outside the region). The index *j* indicates that each relevant organization has to be analysed separately.

[4] Technology management needs for this type of company are analysed in Chapter 3.

[5] It goes without saying that both the breakdown of technology mangement needs and the relative intensities in this Table are to some extent arbitrary. More elaborate versions of the table can be produced and different intensities may be appropriate in different regional or sectoral circumstances. The purpose here is simply to illustrate the approach, not to present a well-developed tool or start a controversy about the appropriate level of disaggregation.

Table 6.2. A representation of company needs

	Large, technology-contingent firms	Small, technology-contingent firms	New technology-based SMEs
Real estate (e.g. BICs, technology parks)		**	***
Technical consulting for troubleshooting (e.g. TT agencies, RAs)	*	**	***
Technical consulting for new products (e.g. HEIs, RAs)	**	***	*
Technical consulting for diversification (e.g. specialized engineering consultants)	***	*	**
Management consulting (e.g. quality control, planning monitoring services)	***	**	***
Human resources development (e.g. HEIs, vocational training, internal continuous education)	***	***	*
Investment grants		**	**
Innovative operations support (scale change)		**	***
Risk capital	*	**	***

Note: * low need,
** medium need,
*** high need.

For the reasons we have discussed extensively above, a reasonable matrix must be well documented empirically and not designed according to the original mission of the RTI organizations concerned. Organizations of the same type do not necessarily offer the same type of services, nor do they offer the same services to all types of client. It goes without saying that in core regions the matrices are better filled than in LFRs. But even in advanced regions, the regional studies suggest that risk capital and consulting services for diversification and growth are not offered or at least not easily accessible to SMEs. Moreover, costs as well as needs have to be taken into account: while the services may be available, their cost may be prohibitive, in particular for technology-contingent SMEs.

It must be stressed that these matrices are only an analytical tool, which will have to be complemented by suggestions on how to fill the gaps they identify. In the rest of this chapter, a presentation of the empirical evidence of our study highlights some good practice examples of how this can be accomplished.

Empirical Evidence for Matching Companies' Needs to the Supply of RTI

In this section we examine empirical evidence gathered in our study and drawn from all types of European regions. As stressed in Chapter 2, it came as a surprise to see that the problems of technology management are more related to the type of firm than to the region. The conclusions of the study are that it is possible to access relevant knowledge from within any region in the EU, but that the propensity of firms to adapt to technological change and to overcome relevant barriers differs.

Although aid to innovation differs substantially from country to country in Europe, cases identified in one region are sometimes easily applicable to other European regions and can be taken into consideration by policy-makers either in creating new infrastructure (filling the gaps in the matrix designed), or redirecting existing service providers to adopt (or avoid) desirable (or undesirable) features.

In the following sections we discuss:

(*a*) *Rhône-Alpes*, a reference 'core region' case study; (Suquet et al, 1993)

(*b*) the 'model' case of *Baden-Württemberg* (Semlinger, 1993; Pyke, 1994);

(*c*) key problems for *new-technology-based firms*, as derived from our case study on Portugal (Coombs and Fontes, 1993), but applicable elsewhere;

(*d*) the relevant features of *larger, globally networked, but technology-contingent enterprises*, resulting from our Case Study on Greece (Tsipouri, 1993); and finally

Table 6.3. Coverage of company needs by each type of organization (j) in a typically advanced region

	HEI_j	PRC_j	RA_j	TT $agency_j$	BIC_j	Technology $park_j$	Technical $consultant_j$	Management $consultant_j$	Banking $institutions_j$	Other financial $institutions_j$	Government $schemes_j$
Supporting NTBF creation	*	*		***	***	**	**	**	**	***	***
R&D support	***	***	*				**				***
Science-based innovation support	***	***	*				*				***
Technology-transfer support	**	**	***	***	**	**	***	**	**		***
Pro-active TT strategy	*		*	**			**				
Quality certification			***	***			***	***			
Prototype construction			***				***				
Support to change scale			**	**				*	**	**	**
Support to extend market			**					***			**
Management consulting			*		***			***			
Human resources development	***	**	*		*						
Investment grants										***	***
Risk capital					*	**				**	**
Information dissemination	***	***	***	***	***	***	***	***	***	***	***

Table 6.4. Coverage of company needs by each type of organization (j) in an average region

	HEI_j	PRC_j	RA_j	TT $agency_j$	BIC_j	Technology $park_j$	Technical $consultant_j$	Management $consultant_j$	Banking $institutions_j$	Other financial $institutions_j$	Government $schemes_j$
Supporting NTBF creation				**	**	*		*	*	*	**
R&D support	**	*	***	*		*	*				**
Science-based innovation support	*	*	*			*					**
Technology-transfer support	*	*	***	**	**	*	*	*			**
Pro-active TT strategy			*	*							
Quality certification			**	**			**	**			*
Prototype construction			**	*							
Support to change scale			*	*	**			*			*
Support to extend market			*			*		**			*
Management consulting					***	**		***			
Human resources development	***	*		*		*	*				
Investment grants									*	*	**
Risk capital										*	**
Information dissemination	***	***	***	***	***	***	***	***	***	***	***

Table 6.5. Coverage of company needs by each type of organization (j) in a less favoured region

	HEI$_j$	PRC$_j$	RA$_j$	TT agency$_j$	BIC$_j$	Technology park$_j$	Technical consultant$_j$	Management consultant$_j$	Banking institutions$_j$	Other financial institutions$_j$	Government schemes$_j$
Supporting NTBF creation					*	*					**
R&D support	*	*	*			*					**
Science-based innovation support	*					*					*
Technology-transfer support			*			*	*				**
Pro-active TT strategy											
Quality certification			*								**
Prototype construction			*								**
Support to change scale								*			
Support to extend market					*			*			
Management consulting								*			
Human resources development	**										*
Investment grants									*		***
Risk capital										*	
Information dissemination	*	*	**		**	**	*	**	**	*	**

(*e*) means to overcome barriers for the *technology-contingent SMEs*, as illustrated by Local Production Communities (Garcia-Blanch, 1993).

Key Features of the Rhône-Alpes Region

The Rhône-Alpes region provides a particularly favourable environment for innovation, with strong incentives and a dynamic industrial structure (Suquet *et al.*, 1993).[6] The twofold driving forces of technological dynamism are scientific capacity in the new industries, and a tradition of innovative experimentation based on accumulated know-how in more established zones and industries. The region is well endowed with large enterprises, with global networks, high internal skills, and a strong R&D potential. These large enterprises co-operate with local SMEs, though they consider them as interchangeable resources: partnerships with SMEs are based on power relations. These firms often deliberately monitor SMEs, especially those located in science parks, in order to identify those that could be developing interesting technologies, with the prospect of eventually acquiring them.

In order to support the region's SMEs in 1989 the Regional Council created the network 'Présence Rhône-Alpes', which groups together the principal technological actors in the region (administrations, laboratories, universities, ANVAR, CNRS, etc.). This network promotes the technological development of SMEs by regular visits from counsellors, who are charged with informing enterprises and sensitizing them to technological opportunities. The present situation is difficult for SMEs; it leads them to concentrate on getting the maximum profitability from their technological investments and to be less concerned with the growth of their capabilities, thus externalizing R&D and neglecting intangible investments. In particular the companies outside the local networks have strong feelings of fragility and see the permanent search for productivity improvement as the major determinant of their survival.

In order not to endanger their long-term survival, and their contribution to the national research system, these enterprises have been identified as a priority for various types of assistance, one of the most interesting initiatives being the creation of companies that

[6] This analysis is based on the Case Study on the Rhône-Alpes carried out in the context of this study.

develop and market university research results. These enterprises are becoming real interfaces and managers of the relations between university laboratories and industry. Their role is to transform industrial demand (after having identified it, i.e. translated latent into expressed demand) and direct it towards the appropriate laboratories, rather than to inform industry directly of the activities of university laboratories, which often do not correspond to industrial expectations.

The conclusion to be drawn from the Rhône-Alpes experience for the RTI is that, even in regions where large companies are present and inter-company linkages play a crucial role, technology-contingent SMEs still need reinforcement. One example of good practice is to support the co-operation of RTI organizations, acting as companies (intermediaries with a business outlook), in order to gain confidence and access the technology-contingent firms.

The German Model Region of Baden-Württemberg

Baden-Württemberg, which is one of the most prosperous regions of the EU, is well known for its dense knowledge infrastructure, which is highly valued by the local business enterprises. The foundations of this model can be traced back to the past century and are based on setting up and using centres of technological competence at different levels, improving the absorptive capacity of firms (awareness creation, training, mobility of personnel), establishing fora for best practice diffusion, supporting the creation of technology-based firms, networking support organizations, and developing international opportunities for accessing science and technology. This generalized strategy has crystallized into three types of strongly interrelated RTI structures each of which addresses the supply side, the demand side, and the problem orientation:

1. A number of industrial research associations, two-thirds of which have their own laboratories, often attached to higher educational institutions (HEIs). These industrial research associations are sector-based, undertake collaborative R&D in response to the needs of SMEs, provide RTD-related services (information, testing, certification etc.), develop industrial networks, and diffuse best practices (based on trust among members).

2. The Fraunhofer Gesellschaft (FhG), which undertakes

applied R&D for industry and the public sector. The FhG is a national organization with an independent corporate identity. It consists of a large number of research establishments, several of which are located in Baden-Württemberg. It has linkages with the technical universities and one-third of its funding comes from the public sector (*Bund* plus *Länder*). The FhG is organized around technologies, not sectors.

3. The Steinbeiß Foundation provides technological support for SMEs in Baden-Württemberg. It is highly decentralized with a network of about 1,000 Steinbeiß-associated consultants, making use of the competences of technical universities. The working mechanisms are basically demand-led.

This set of structures covers all dimensions of company needs and provides a wide basis for intermediary functions: information, locating and assessing resources, bringing together supply and demand, developing structural arrangements, monitoring transfer and use of new technologies, and adapting technologies to needs. This variety of structures is effective because the linkages between the providers, intermediaries, and users of knowledge are very intense. It is the synergies in the system that allow all elements of demand to be covered. At the same time the supply-side benefits from a climate of confidence which allows technology-push strategies to be effective. Thus it is not merely the existence of a dense knowledge infrastructure but the interaction between the elements of the system that improves technology management. Co-operation, competition, overlapping patterns, and even conflicts among the RTI organizations can have a creative role, as long as their administrative and financial autonomy are guaranteed.

Small, Technology-Intensive Firms: Some Evidence from Portuguese NTBFs

NTBFs are an interesting group in Portugal; by virtue of their attitude and behaviour they contrast strongly with the structural inadequacies of industry in the Less Favoured Regions (LFRs). Despite their relative immaturity, NTBFs may have some potential to help modify the prevailing bleak scenario in these regions. The group of companies investigated in our case study in Portugal (Coombs and Fontes, 1993), with relatively low turnover and exporting activities, did not conform exactly to the classical definition of NTBFs, as it included a few cases of young firms that despite their technological orientation could not be accurately described as NTBFs, and a few cases of young technology-based firms launched by existing companies. The latter, together with some NTBFs recently acquired by large firms, may provide a window on some of the technology strategies of those larger firms. These firms see their most important objectives as increasing their market share and creating new markets, as well as the extension of their product range. Most of them were originally established as 'one-product firms'—corresponding to the tendency towards over-specialization discussed in Chapter 4, which can be found everywhere in Europe. The strategies followed by these firms are bounded by the small size of the Portuguese market, as well as by the image of the country as a low technology and low cost producer. This environment combined with a limited availability of knowledge locally and an inadequate transfer of this knowledge to industry explain why in general Portuguese NTBFs cannot aspire to have strategies with a radical technological content. Offensive strategies are beyond their reach and defensive strategies can only be envisaged by few firms with very particular characteristics, including some firms strongly linked to research institutions; thus most of these firms have imitative strategies, based strongly on their capacity to master existing technologies in very specific areas. Technology and business strategies are closely interwoven, since frequently the strategy is as much based on an appreciation of the market as on the relatively narrow set of technological competences available to the firm. Strategy is often implicit and adaptive so as to accommodate to a small market, in which over-specialization is dangerous.

Firms' activities are inevitably founded on a small number of core competences and technologies that are the basis of their products and services, but they have different approaches to developing these competences. Some firms simply consolidate their existing capabilities, others explore widely the opportunities these competences provide, upgrading them and monitoring relevant scientific and technological developments; only a small group has a longer range perspective, trying to identify the technologies that might become relevant in the future and working to gain competence in them, in parallel with their present activities.

Portuguese NTBFs organize their technology management to the largest extent around product development, in some cases with in-house research activities, but frequently with no separation between research and engineering staff. External acquisition of technology is important, always in the form of informal contacts and in specific cases involving formal relations with the technology infrastructure in the country. Most firms stated that, when developing new products, ideas coming from outside are more important than own ideas. To be more precise, internal ideas emerge either as a response to a specific client solicitation or after the identification of a client's need. Suppliers—because they sometimes offer new solutions—and competitors are also important in this process. The risk of introducing a product in the market only on the basis of the firm's own initiative is too high, and this limits the strategic product development options.

The single most important problem faced by the more technology-orientated firms, and most commonly identified as one of the greatest obstacles to an increase in the technological level of Portuguese industry, was the shortage of skills. Skilled workers are usually difficult to obtain in all areas and young people often do not have technical skills at all, due to the interruption of vocational training in Portugal. Despite the number of short training courses launched in recent years, firms continue to find it very difficult to find skilled workers and have to provide in-house training for the workers they recruit. The problem is more acute than in the rest of the economy because experienced people do not generally take the risk of joining a small and unknown firm, especially in its early years.

In order to improve technology management in this type of company long-term changes have to be triggered, to increase technological awareness, to change the country's image and to move away from competitive advantages based on cheap labour. At a more concrete policy level, firms need to be encouraged to search externally for complementary competences (associations, networking), and search for these competences beyond the regional context; the improvement of vocational training and higher education in the service of industrial employment is also needed. The relevant contributions from the RTI then relate mainly to guidance towards new technological abilities, sensitization to next-generation technologies, and to the stimulation of networking, when it does not occur naturally.

The Bigger, Globally Networked Greek Companies

The relatively small number of bigger companies in Greece (which according to EU definitions still means SMEs since often their staff is only 200–500 employees) demonstrate a very rapidly growing momentum towards international networking, originally provoked through the need to identify sources of technology which were lacking in the country, but also supported by the massive new incentives for transnational collaborations such as Community R&D funds (Tsipouri, 1993). Companies of this type have well-elaborated business strategies and their objectives in promoting technological change are mainly to extend product range and to lower production cost. The latter tends to be associated with investment in new vintages of capital equipment rather than the improvement of in-house technological competence or the conclusion of licensing agreements. In that sense they may be considered as cases of large, technology-contingent companies. Overall licensing and patenting is very limited, far from the general business custom. A characteristic of the technology management of these companies is that, although they have a top management strongly committed to the need to adapt to technological change, they regard this as an exogenous process, and base their decisions on information provided by foreign suppliers of equipment and raw materials, or well-established engineering and consulting companies (again mainly from outside the country; the same low skill and low cost image that inhibits exports of NTBFs in Portugal acts against local interaction in Greece). They combine external advice with good in-house training in the application and use of new technology.

More technology-aware than the smaller SMEs, the big, technology-contingent firms seek to keep pace with international technological developments either through contacts outside the country or by fragmented actions often responding to public policy incentives. They have a good technological competence in terms of monitoring evolution, but they are less offensive than NTBFs. Their entrepreneurial competencies and learning abilities are better than those of technology-intensive SMEs, but they have strong needs for diversification strategies and the constant examination of possibilities for upgrading their competitive position, through the creation of new directions and skills and by passing from defensive to offensive strategies. For them the local RTI can be valuable, in identifying new technological opportunities related to local circumstances and possibly for reducing the cost of access to new knowledge.

These bigger companies, which are rather successful in both national and export markets thus seem to be suffering from an 'invented-elsewhere' syndrome. Although this pattern was discovered in our case study on Greece, it seems also to apply to a considerable number of big, technology-contingent companies all over Europe. Such companies tend to regard the active pursuit of technology as a waste of time; although they realize that the introduction of technological change is an important element for their growth, they do not believe it makes economic sense to organize R&D departments or to hire people who will devote all their time and energy to the creation of new knowledge; nor do they envisage gaining income from active licensing. In these firms engineering and quality-control departments are well organized, while R&D departments, if they exist, are very small. The whole process of adaptation to technological change is supervised by the general manager. Beside the risks to the company itself that arise from the absence of any effort to develop learning abilities, such an approach represents a strong disadvantage for the regional or national economy, as the most successful companies are not playing the role of 'champions', which can be used as leverage for creating or upgrading supply chains. Moreover, although they would be expected to present the largest element of demand for local infrastructure services, in fact they hardly acknowledge the availability of these services. This invented-elsewhere syndrome seems to be characteristic of all average skills companies be they locally or globally networked. Some globally networked high skills companies are also affected by the same syndrome, in particular in areas outside their core competences. High skill and locally networked firms (mainly NTBFs) are the only category which clearly escape it.

The Majority of Small, Technology-Contingent Industrial Firms and the Local Production Communities in Spain

The majority of firms in LFRs are, as mentioned above, neither NTBFs nor bigger companies with some technology awareness. They employ less than 100 people, are faced with considerable business threats resulting from financial inadequacies or increasing competitive pressures, and operate in traditional sectors, where they see technological change as a luxury which does not concern them directly. While it is generally very difficult to change the attitude of this type of firms so as to explicitly incorporate a technology strategy, the case of the Local Production Communities (LPCs) in Spain seems an interesting model to study (Garcia-Blanch, 1993). High growth rates since 1985 have attracted many multinational companies to compete in Spain. This has resulted in increased competition in almost all market sectors, squeezing out local companies. The strategic problem for most local firms now consists in diversifying their product portfolio at very low costs to compete on the Spanish market and at the same time to improve the quality and external features of products, such as design, presentation, packaging etc. in order to have some prospects in external markets. A Local Production Community is defined as a set of industrial companies in a specific geographical environment, where certain types of co-operative relations frequently based on personal interaction have developed, which boost each company's production and also that of the local economy.

The productive process is divided over a network of autonomous production centres, each company concentrating on specific production tasks. Decentralization in an industrial branch takes the form of process specialization. Co-operation in the phase of research, manufacturing, and commercialization among different companies emerges as a result of specialization and the complementary activities of each firm. In such circumstances geographical proximity can be considered as an integrator of the externalities of production. Analysis of the industrial organization of the LPCs allows us to consider the linkages between efficiency and competition within a particular geographical area. Collaboration between the companies involved generates synergies, and accelerates innovation processes. And it is believed that openness to influences from the international market improves the competitive position of local businesses within the LPC.

In general, one can state that this is a model case of a joint group with a positive attitude towards innovation, effectively exploiting the dynamic potential peculiar to the LPCs, which is based on personal relationships with the neighbourhood, and which lacks structures that can condition or limit future perspectives. A lot of emphasis is placed on design, but not very many external services are used. Likewise, the number of product innovations being carried out is quite high, partly because they are motivated by pressure from the competition. However the associated companies do not have formal research and development departments and the number of qualified technicians employed is quite low. Neither do they have any agreements with external research and development centres. The sources of information which are neither very formal nor systematic reflect personal ties with other companies in the area, and the importance of client and supplier relations, either from within the area or from outside.

The example of LPCs suggests a relatively successful (even if not explicit) management of technology, despite a near absence of formal R&D activities, a virtual absence of collaboration agreements with external investigations centres, but modern equipment and a clearly favourable attitude towards technological innovation. It seems that the strength of this common agreement lies in trust, which helps companies overcome cultural barriers. As a conclusion we argue that one major element that distinguishes the LPCs from other technology-contingent firms in LFRs is that within the framework of their collaboration they offer each other support which resembles a private infrastructure, used without any fees. One may then see their success as based to some extent on a substitute for RTI, which has the advantage of being highly flexible and confidence-generating, as it respects the principle that 'companies learn best from other companies'.

Adapting Research and Technology Infrastructure to Technology Management Needs

It is suggested in this chapter that the role of the RTI in supporting technology management is very important, and is becoming more important and more challenging over time. There are several reasons for this.

1. *Technology is becoming constantly more complex and pervasive* and changing at a rapidly increasing speed. As a consequence companies cannot ignore the need to accommodate technological change, but as not all firms have the means to monitor technological change internally, external support becomes inevitable.

2. *A competitive company accommodates technological change not just through internalization of new knowledge*, but through a

constant improvement of its technological and entrepreneurial capabilities, and its learning abilities. Thus the role of the RTI becomes more challenging, as it goes well beyond the need to monitor and diffuse scientific knowledge: the RTI needs to complement company competences in a rapidly changing world.

3. *Technology-contingent SMEs in particular, but bigger companies as well, need to rationalize in order to survive* and this occurs to a large extent at the expense of intangible investments in R&D, technology transfer, and training. The RTI must then be ready to support these companies to successfully externalize these functions, which they can no longer afford to maintain internally.

4. *Finally, the RTI can, if adequately designed and managed, play a substantial role in covering the technological and entrepreneurial needs of firms*, in the face of market failures and insufficient linkages between companies.

But although the need is theoretically well founded, in the real world the RTI does not always meet the challenges. In the 'model' regions of Europe (e.g. Baden-Württemberg), where technology infrastructure is credited as an important contribution to company success, not only does it help companies with expressed needs to turn knowledge into useful products and processes, it also manages to transform latent demand from technology-contingent companies into actual demand. In these regions the whole range of technology creation and intermediary functions is covered. This variety of structures is effective because the linkages between providers, intermediaries, and users of knowledge are very intense. It is the synergies in the system that allow all aspects of demand to be covered, and create the climate of confidence which benefits the supply side. Consulting companies and RTI organizations are able to operate on break-even principles or even a profit-oriented basis and gain the confidence of firms, because, as stressed in our basic ideas 'companies learn best from other companies'.[7] Thus it is not merely the existence of a dense knowledge infrastructure but the interaction between the elements of the system that offers firms better opportunities to improve their technology management. Co-operation, competition, overlapping patterns, and even conflicts among the

[7] The Danish Technology Institute was found to be a particularly interesting case; when obliged by the Danish government to rationalize (work more like a company in our terminology), it actually increased its turnover and number of contacts with companies.

organizations involved can have a creative role, as long as their administrative and financial autonomy are guaranteed. A first task is then to make sure that all types of support are available or at least easily accessible. But, since the model cases are the exception rather than the rule, it is important for policy-makers to have available both analytical tools and reference cases[8] in order to design their RTI strategies. The case of the Less Favoured Regions in the EU shows that merely creating additional infrastructure is insufficient.

The insight that technology management is more than just the acquisition of technological competence is the key angle under which we discuss the creation and management of the RTI. The performance of an enterprise depends not only on its own competences, but also on the competences available in its environment. The point is not that competences are nurtured by the knowledge infrastructure (e.g. by the education of future employees and by research carried out in universities), but also that the uses an enterprise makes of its competences depend on the challenges imposed by the environment. Traditionally, the knowledge infrastructure is seen as a mechanism to improve technological competence alone. In that respect it is by definition limited in its role. Unless the RTI takes the initiative to directly support (through management training, demonstrations, etc.) the improvement of entrepreneurial capabilities and learning abilities, its contribution to technology management and competitiveness remains limited. Successful RTI organizations should be taking the initiative to become pro-active shapers of strategy support, rather than just technology providers.

To implement this support strategy firm characteristics are the most important common denominator for designing the approach. Small technology intensive enterprises co-operate well with universities and research centres, but their needs go beyond what this type of knowledge infrastructure can typically offer as has been noted in Chapter 4. For them our study has shown that assistance in areas such as business and marketing plans is more relevant. Large technology-intensive enterprises which compete on the basis of technological competence also tend to have strong linkages to local and trans-European RTI, but divergent cultures and competition for intellectual property rights limit the scope for co-operation.

[8] These should be considered not as 'best practices', but as guidance and incitation to policy.

These companies have the culture and the means to further intensify their linkages with knowledge-providing institutions and need to fight against their own inertia and their characteristic not-invented-here syndrome. Large and small technology-contingent enterprises, those that in principle have the highest need to adapt to technological change, have fewer linkages, partly because they lack the qualified personnel that are able to communicate with public organizations. They constitute the main challenge for the RTI, which ideally has to build a climate of confidence, turn latent into effective demand, and cover the entire scope of demand from technological needs to the improvement of learning abilities. It is in order to respond to this challenge that policy-makers need to carefully examine the existing supply of services of the RTI and see what aspects of the total spectrum of demand they really cover. The educational system, as part of the knowledge infrastructure, is the primary source of qualified labour and is important to all enterprises, particularly those that are usually at the end of the line in the labour-market. Technology transfer infrastructure is mainly of use to SMEs that have already been able to build up some in-house technological competence. The impression is nevertheless that some of these transfer organizations are failing to reach the customers for which they have been created.

Whilst the needs for technology management are basically determined through firm characteristics and behaviour, sectoral considerations should also be taken into account. The specialist supplier sectors, i.e. sectors with low concentration and product-focus or technology-producing firms (see the case studies in the software and mechanical engineering sector described in Chapter 4) have in-house and sometimes too specialized knowledge. For them access to knowledge infrastructure is necessary both for skilled personnel and in order to follow and internalize technological developments outside their field of specialization. The science-based sectors, like chemicals and electronics, are characterized by relatively high levels of concentration and large, multinational enterprises with sizeable R&D activities. In all science-based industries firms are trying to find ways to organize and control the flow of information coming from the outside concerning the wide range of technologies that may be of potential importance to their business. In science-based industries basic research is an important component, so long-standing relations with universities and research establishments are important. Here, the particular power of public bodies to impose regulations can give rise to new (niche) markets, but most enterprises experience regulation as an obstacle to innovation. In the supplier-dominated sectors (like food processing, clothing, and construction), technological change is mainly produced by the suppliers of materials and equipment. Their base technologies are well known and can be implemented everywhere. Innovations are incremental. Although supply of relevant knowledge may have been a bottleneck in some regions and some sectors, the real bottleneck now is that many firms and their owners lack the capability to make use of the various sources of knowledge available to them. Small enterprises tend to be more reliant on in-house sources of information than large enterprises: the capacity to do research and the capacity to absorb knowledge from the outside are closely related. Finally the scale-intensive sectors (aerospace and car manufacturing) are highly concentrated and are typically those with the best collaboration patterns with knowledge infrastructure.

7

Conclusion: *Technology Management in the European Union and its Policy Implications*

William Cannell

Introduction

The three components of firms' technology management capabilities posited in our study—technological competence, entrepreneurial competence, and learning ability—have provided a platform from which to explore the diverse and complex elements involved in the practice of technology management. We have shown that firms' capabilities in managing technology are indeed a function of their management as well as their technological abilities, and that a fourfold typology of firms provides a good basis for identifying the major types of problems of technology management they confront.

In the earlier chapters we have explored the key issues arising in relation to specific types of enterprises, and reviewed the interactions between technology management on the part of firms and the research and technology infrastructure. These chapters have addressed in some detail the main policy questions raised in the first chapter and provided specific recommendations for policy. The aim of this final chapter is to take some of the important findings and strands of argument presented in previous chapters, slightly further to situate them in the context of other literature and thereby to provide more general guidance for policy and the role of the public authorities.

The views expressed in the chapter are those of the author and should not be taken to represent the position of the European Commission.

The Nature of Technology-Management Competence

From our perspective technology management is about developing and maintaining capabilities to ensure that in the short and long term technology can be harnessed to business goals. This perspective reflects our concern with policy. We wish to distinguish the generic capabilities that underpin technology management, and which are thus a legitimate concern of policy, from the strategic dimension of technology management in firms, which is proper to the management of companies themselves. Our results indicate that it is indeed possible to do this. Hence, it is plausible that policies can be developed to improve technology-management capabilities without compromising the capacity of firms to generate the diversity in technology management approaches which is needed to maintain a dynamic and innovative economy.

What firms can aspire to in the future, in technological and commercial terms, depends on the competences they possess at present. The path-dependent nature of technological learning (David, 1985; Nelson, 1992)—is amply illustrated by our sectoral case studies. Technology management in individual firms, whether implicit or explicit, formal or informal, is intimately tied to the particular circumstances of the firm in question. Our four-way classification of firms addresses the main elements of variety in technology management; however at a more detailed level the specific character of technology management in individual firms is clearly a function of their business circumstances: market position (market niches and product lines) specific technological capabilities (product technology portfolio, process technologies), established external linkages and networks, strategic orientation (i.e. offensive, defensive, imitative) etc. The style and organization of technology management and the decisions involved, thus contribute strongly to, and are in turn shaped by, a firm's differentiation with respect to competitors.

But technological competences are by no means the only consideration; maximizing firms' strategic opportunities appears to depend upon a fine balance between technological and broader management capabilities. At very low levels of technological capability the very concept of technology management is necessarily alien to firms, and technology management is implicit at best. Developing more sophisticated forms of technology management depends in such cases on improving technological capabilities. However, at higher levels of technological sophistication this no longer seems to be the case. We find that, on the whole, technology-intensive firms are most lacking not in technology but in the organizational and managerial capabilities to make the most of their technological assets.

Technology-management competences are embodied in individuals and organizational routines relating to technology at different levels in firms. By their nature they are decentralized. As we have seen, this poses problems of co-ordination, especially in large firms, which conduct a permanent search for the optimal balance between divisional and corporate responsibilities for technology and the improvement of interdivisional communications. Moreover, the 'unit of organization' in technology management increasingly transcends the boundary of the firm, which poses many concrete problems of interface management. Product technologies may need to be co-ordinated through standardization at various levels of formality. Process technologies, product design, engineering, and development may need to be co-ordinated with other firms along the supply chain. R&D may need to be conducted partly to accommodate their specific inputs. Some firms may be in a position where they have to respond passively to the strategy of a larger customer. In other cases—joint technology ventures between firms at the leading edge, shared design and development activities along supply chains, or partnerships between SMEs and public-sector support agencies—strategies may have to be co-ordinated in a more 'symmetric' fashion.

It is also, however, clear from our study that the technological and organizational dimensions of technology management competence are in practice difficult to disentangle. Learning involves messy interactions between different types of knowledge and behaviour (Rosenberg, 1991; Nelson, 1992) and is a product of experience and experimentation—'doing', 'using', 'failing', and learning from competitors for example (Pavitt, 1992)—as well as formal education and training. Its takes a form which is highly specific to the individual firm, its strategy, organization, and assets. This 'private' context of knowledge development means that the circulation of 'generic' knowledge and know-how between public and private realms is an important aspect of technology policies.

Commercial, Technological, and Cognitive Elements of Technology Management

Taken together, the analyses presented in the earlier chapters suggest that technology management involves a conjunction at the level of the firm of considerations arising in three different spheres—scientific/technological, commercial/economic, and cognitive/organizational. Managing technology in the firm, and developing policies to improve technology management are both a matter of coming to grips with the distinctive logic of each of these spheres, and the diverse interactions between them.

The Technological Sphere: Complexity and Science-Intensity

Technological advance and the broadening and deepening of knowledge on which it depends naturally lead to a process of increasing specialization, of research units, firms, even regions and nations (Archibugi and Pianta, 1992). The consequence of this is an ever-tightening web of relationships between different carriers of knowledge and expertise, both vertical (leading from the creators of knowledge such as universities downstream to private-sector users and consumers) and horizontal (between different strands of knowledge or different technology-based elements of products and systems). In the earlier chapters, several examples were given of the ways in which these effects present themselves to firms and confront them with both strategic and organizational problems. The forces at work are themselves a function of several factors, including the level of maturity of the technologies and markets involved.

Technological diversity in mature products is outstripping functional diversity. The dominant design configurations of mature products, such as aircraft, cars, petrochemical plants, or machine tools, tend to be quite stable, constrained by the enormous investments made in accumulating capability and associated infrastructure, standards, and so on. Large supply networks are involved in the manufacture of such products. As noted in Chapter 3, the technological diversity within these systems—the diversity of technologies making up the components of the overall system—is growing in step with growth in performance and/or overall functionality.

Already-complex systems are being integrated at a higher

functional level. The process of system integration raises complex optimization questions as well as standardization and interface issues, for both manufacturers and users, as for example in the machine-tools industry where 'product business' of stand-alone equipment is being transformed into 'systems business' (Reger and Kungl, 1993). The combination of computer-aided design (CAD) with computer-aided planning (CAP), and computer-aided manufacturing (CAM) is being transformed into computer-integrated manufacturing (CIM). Fundamental problems arise for suppliers of defining the overall architecture and functionality of such systems in a situation of great uncertainty and with little opportunity to experiment, and for users of accommodating them into their business practices.

Innovation is occurring at the interface between different strands of technology. This includes the phenomenon of technology fusion, and observers (OECD, 1993*a*; Gibbons *et al.*, 1994; Kodama, 1995) suggest that it provides the major source of opportunity for technological innovation in the near future at least. Here, one confronts a variety of problems, illustrated in Chapter 4, associated with searching out and bringing together productive combinations of different scientific disciplines and technological practices, and creating commercial and organizational arrangements which allow competence to accumulate without compromising progress in the separate underlying bases of knowledge and expertise.

The links between product and process innovation are growing. Shorter product life-cycles and accelerated development times mean that developing and optimizing process technology has to be done concurrently with product development. As products themselves become more complex, performance standards in areas such as quality and reliability become increasingly dependent on production standards. Process innovations are also crucial in moving towards strategies of flexibility and mass customization. Finally, instrumentation and information management technologies are now a critical part of research and innovation management in high-tech industries.

Technologies are becoming more closely dependent on the science base. This is shifting the boundaries between firms and public providers of science and technology, specifically the university system. The increasing market relevance of fundamental research areas like information science, genetics, and solid state physics, combined with budgetary constraints is making university departments much more eager to become commercial players in their own right. As noted in Chapter 4, it puts at issue some of the traditional thinking on the university–business interface and the role of bridging institutions.

Overall, it appears that the opportunities for technology development presented to firms have widened considerably, with two major consequences. First, firms need to conduct broader searches for technological inputs both internally and externally, including increasingly in areas where their proficiency is low, and to develop broader partnerships with various sorts of technology providers. Second, firms—individually and in networks—must become increasingly concerned with the integration of technologies into products, processes, and along supply chains.

The Cognitive Sphere: Cognitive Limits of Individuals and Organizations

The increasing complexity and speed of change of scientific and technological knowledge, rising competition and market demands, and diminishing product-development cycles, place serious cognitive demands on innovation and research actors. Innovation and technological change manifestly requires knowledge and understanding. Although this can be codified to some extent and/or 'black-boxed'—embodied in tools and equipment the operation of which does not need to be understood by the users—mastery of technology, as well as the other competences involved in technology management, remains to a great extent a function of the tacit knowledge of individuals (Nelson, 1992). Organizations like firms can be considered as mechanisms to amplify the cognitive capacity of individuals; the same can be said, at a lower level, for teams of individuals within firms or, at a higher level, industrial systems, indeed societies as a whole (Lundvall, 1992).

A firm's cognitive resources influence the attention it can give to strategic and operational demands, and underly what we have called 'learning ability'. It is clear that they are a complex function of workforce numbers, intelligence, skills and motivation, organizational structures and routines, information systems, external linkages, and management philosophy. The dynamics of learning, and the implications of choices regarding the ways in which cognitive resources are allocated to the various aspects of accumulating, storing, and exploiting knowledge, are important 'boundary conditions' for technology management. At the same time they create behavioural regularities, associated with 'bounded rationality' (March and Simon, 1993) on the part of firms, which are essential for policy-makers to take into account.

The limited absorptive capacity of firms for technology, though particularly evident in our examination of small firms, must be regarded as a general phenomenon. Firms are able to identify and assimilate technological know-how to the extent that they already have technological (and entrepreneurial) competences in the same fields. The speed with which firms are able to learn about new developments in technologies, applications, and markets, depends on their existing level of competence. Without some technological competence, no learning is effectively possible; as technological competences increase, so does the capacity to learn more. Learning, in short, is cumulative (Nelson and Winter, 1982; Dosi, 1988). The directions in which firms are able to learn depend on the types of competences they have already developed. The search for technological opportunities is normally conducted within the perspective offered by existing technological competence. Small, technology-intensive firms, as we have noted in Chapter 5, tend to search around a single technological trajectory. Even large, multi-technology firms have been found to search for and exploit opportunities which reflect their existing product mix, so that their technological profile remains very stable over time (Patel and Pavitt, 1994).

Organizational development and human resource policies, as well as research activities, clearly provide scope to improve cognitive resources. Training programmes (e.g. directed towards multi-skilling), new organizational paradigms (such as lean enterprise, decentralized management, flatter hierarchies), new organizational routines (e.g. based on process re-engineering, or total quality management), improved external networking, and so on are intangible investments which can have quite dramatic effects on a firm's ability to accumulate and exploit technological knowhow. But like other investments they involve financial commitments, risks, and opportunity costs. The allocation of cognitive resources demands several types of choice.

Between investment in technological capabilities, and those involved in related activities such as production or marketing. As we have observed, technology management competence does not reside in technological competence alone. Mastery of technology in a commercial context involves an integrated set of technology and market-related capabilities which are developed and maintained through constant practice.

Between corporate and divisional interests. Firms cannot optimize their learning ability at all levels; we see this difference being worked through in the progressive development of 'paradigms' of technology management especially in large firms (Rothwell, 1992; Coombs and Richards, 1993*a*).

Between the long and short term. Even the most rudimentary form of technology management requires resources to be diverted from everyday management activities towards developing competences of interest beyond the immediate term. A constant practical problem for firms in designing their management systems and business processes is how to ensure the proper balance between static efficiency and organizational learning, the latter contributing to longer-term adaptation but making demands on current resources.

Between internal and external capabilities. Building links with suppliers of technology and related expertise can involve considerable demands on internal resources: internal and external learning are complementary not substitutes. This has the paradoxical result, noted in Chapters 5 and 6, that large firms tend to be more reliant on external sources of knowledge than smaller ones.

The essential danger for firms arising from limits to their cognitive resources is clear: missing potential innovation opportunities as a result of limited absorptive capacity and inadequate scope in search processes. The problem for policy is that these limitations are systematic—each and every firm finds technological learning easier as it develops technological competences and easier in the areas it already understands. The likely results include a vicious circle of suppressed supply-and-demand for technology, amongst technology-contingent firms, and sub-optimal levels of innovation through technology integration—an area which we have earlier suggested is particularly opportunity-rich—amongst the rest.

The Commercial Sphere: Competition, Markets, and Technology-Management Responses

If it is to be supportive to business, technology management must be driven by commercial needs. The competitive environment appears in our study as a crucial incentive to the development and improvement of technology-management capabilities. The preceding chapters have consistently illustrated how competitive forces, arising especially from the process of European economic integration, have been the main instigation behind the recognition of technology management as an important competitive asset. Competitive challenges, actual or anticipated, have in most cases lain behind the recognition by firms of the importance of technology and fostered the improvement of technology management capabilities.

The manner by which competitive forces come to be focused onto technology management is of great importance. Industrial and market organization are crucial factors, in so far as they impose demands for lower cost, higher quality, and other improvements in product performance via customers and/or clients—mechanisms which have been extensively discussed, for example by Porter (1990), Lundvall (1992), and Von Hippel (1988). Commercial relationships are structured by, and in turn themselves influence, the shape of the technology-based products to which they contribute. Products such as cars embody a hierarchial 'architecture' which corresponds to the logic of problem-solving in design, engineering, and manufacturing processes, each requiring contributions which incorporate a large number of technical disciplines, production operations, and individual companies whilst at the same time providing the basis for a product 'concept' to guide consumers choices (Clark, 1985; Meyer and Utterback, 1993). Our Case Study on Mechanical Engineering indicated that a similar 'modular' concept is developing in the machine-tools sector (Reger and Kungl, 1993). As the technologies embedded in these products evolve and increase in complexity, this threatens established commercial relationships between the firms involved. Even in these relatively mature sectors, it is the technologically more advanced and adaptable firms that stand to gain from this process. Geographical proximity is another important influence on the type and strength of commercial relationships; our Case Study on Spain, as noted in the previous chapter, illustrates this clearly for the particular circumstances of Local Production Communities (LPCs).

What our study shows is that internal mechanisms—planned or not—are also vital in generating and focusing innovative pressure within the firm, in transmitting these pressures to suppliers of technologies and technology-based components and services, and in defining a productive 'agenda for choice' (Metcalfe, 1995) regarding technology in the firm. They include the existence of marketing functions of a satisfactory level of technological sophistication to identify or formulate product opportunities, the channels between marketing and development which are needed to translate functional and operational demands into technological demands, and the interconnections between research, development, and other departments necessary to put the technological and related knowledge together into product and business development options.[1] Of the many barriers to such investments, the lack of internal management resources and external funds are of particular importance. As we have seen, technology management inevitably requires resources to be diverted from immediate concerns to developing competences of longer-term benefit. Building links with suppliers of technology and related expertise can also involve considerable demands on internal resources.

Also important, it seems, are the perceptions of risk and likely reward on the part of management, which demonstrate characteristic patterns across our different types of firms. For small traditional firms without any existing internal technological competence, a significant element of risk arises in their inability to evaluate technological options available to them, either internally or through the medium of external advisors, in the latter case because they have not built up the necessary trust which arises from close and long-term business relationships. For technology-intensive firms, particularly smaller ones, a major element of risk arises in the indeterminacy of market outcomes regarding competing technological options and the possibility of backing the wrong technological horse.

Our results tend to suggest that for the more technology-intensive firms upgrading technology-related capabilities is connected with a geographical widening of market and technology

[1] Our fourfold typology can be useful as a tool for firms to analyse their strategies in those areas, see CEST (1995).

links, as companies focus their product strategies and search for markets further afield or attempt to upgrade or diversify their products using new technological inputs. In doing this, firms may need to radically rethink their business and market strategy, reposition themselves with respect to other players in the industry, and adapt culturally to the new circumstances. An example is the attempted transition being made by a number of chemical firms in Europe away from commodities to fine chemicals and specialities (Cobbenhagen and Philips, 1993). These firms are moving from a scale-intensive to specialized supplier context, having to adjust simultaneously to a dramatically more entrepreneurial environment, batch rather than process operations, a far wider constituency of suppliers and customers, etc.

Upgrading Technology Management: Its Links with Industrial and Regional Development

Our results show clearly that if policy is to address the problems of technology management it must do so within the broader logic of industrial development, and maintain a balance between the different types of possible intervention, according to the existing capabilities of the constituency of firms which is being addressed. Technology management competence is a consequence of industrial development and, conversely, contributes to it, most directly through the competitiveness of individual firms. But active technology management, on the part of large firms and technology-intensive small ones, also provides a form of leverage—through supply chains, the RTI, and by example—over the capabilities of other firms. Companies themselves form a major part of the research and technology infrastructure.

As we noted in Chapter 6, a progressive approach to the improvement of technology management competence is needed, of which the infrastructure is an important part. The balance of factors, not the absolute level of any individual resource, is the main consideration. This requires a sensitive understanding of the particularities of the situation at hand, and an effectively decentralized approach. Examples of best practice which have worked well in particular circumstances can be an important contribution to policy development, but they need to be adapted and implemented with caution. European structural policies have been responsible for considerable investments in the research and technology infrastructure, especially in the less developed parts of the Union. But more infrastructure, it seems, is no longer the solution; pushing the supply side in the absence of sufficient expressed demand may simply exacerbate existing disparities. And the problem of making explicit latent demand has yet to be solved, a matter which is only now being examined in detail.

A general lesson arising from the analysis presented in the earlier chapters is that 'companies learn best from other companies'. The implications of this observation have been considered for the development of the RTI, in a regional context, in Chapter 6. It presents a policy dilemma of considerable significance. Policy-makers must recognize that the effectiveness of their actions in designing and implementing the research and technology infrastructure depends on providing the organizations constituting the infrastructure with a business-like outlook, and allowing them to respond to 'market' demands. But market-orientated strategies lead almost inevitably to a mutation of the role of such organizations away from the objectives they were originally given, towards serving the most attainable, or attractive, market prospects. Such a process tends to undermine the original policy intention, which is usually to fill gaps in the provision of services. The unfortunate effect is to feed economic divergence, with the most able infrastructural organizations in the less favoured regions moving towards serving the most rewarding clients—in both financial and technological terms, e.g. forming international links in preference to local ones. This leads to the risk of a 'dual economy' which does little to foster—indeed works against—local economic development.

The Role of the Public Authorities in Supporting the Development of Technology-Management Competence

We do not suggest that there should be a 'policy' for technology management, or that it should constitute the sole basis for technology policy. Although we have argued that firms are mainly responsible for technological change and that technology management in firms has remained a 'black box' for too long, this does not mean it is the only, or the most important, consideration in developing technology policies. From an evolutionary perspective we must consider technology management as one significant contributor to the 'comprehensive innovation process' highlighted in Chapter 1 (OECD, 1988*a*), a process which also depends for example on long-term investments in basic research. Our approach has therefore been neither to explore all aspects of policy bearing on technological change nor to construct a new policy system, but to use the knowledge we have generated to improve or fill gaps in policies, to highlight generic policy problems and to propose specific measures. The analysis contained in earlier chapters has indicated policy needs arising with respect to each of our four types of firm, as well as the research and technology infrastructure. Taken together and considered in the light of the broader analysis above, these policy conclusions can be presented in the form of a more general philosophy regarding the role and broad objectives of policy in supporting technology management in firms.

Policy should provide supportive conditions for the development of technology-management competences but the application of these competences is a matter for firms themselves. It is in the nature of technology management that it should be closely tied to the strategic objectives of firms, a matter which is definitively outside the realm of policy. As we have noted, however, technology development involves generic capabilities and relies on basic resources which can be regarded as legitimate public concerns. It is this distinction which must guide policy development and implementation, but the practical difficulties should not be underestimated, since competences grow through application in practice. However, the frontier between public and private knowledge is continually moving—by definition—as our general understanding of technology management improves. The firm-specific approaches to technology management which have been found in the past to be most successful are the basis for today's promotional measures on the part of government.

Policy should work at different levels to provide progressive and co-ordinated encouragement to the emergence of technology-management competences, with the help of market forces. The first and incontestable level of intervention is that of basic resources: provisions for research, education, and training facilities, and the availability of finance under appropriate conditions. The second level relates to promotional measures of various kinds, including research, demonstration, and innovation programmes targeted at the broad range of competences associated with technology management. The third level involves easing the conditions for effective interaction and self-organization of firms and other participants in the system of actors involved in industrial technology: ensuring macroeconomic and institutional conditions—including competition rules, standards, and intellectual property regimes—which are favourable to innovation and growth in firms; maintaining institutions and infrastructures dedicated to promoting the joint production of knowledge and expertise and the flow of knowledge and expertise between actors; and maintaining open channels of access to networks by new actors with diverse and useful complementary expertise. By this combination of means to improve self-organization and self-awareness and the institutional changes it implies, the ultimate objective is to improve the capacity of enterprises and the industrial system as a whole to adjust to structural change (Justman and Teubal, 1995).

Policy should be applied non-sectorally but can be delivered through sectoral links. There is no a priori reason why particular sectors should be targeted by measures directed towards improving technology management in firms. Our results suggest that the performance of industrial enterprise could be improved in all the sectors we have studied through a more integrated approach to technology management, and that the main problems facing firms are not sector-specific. On the other hand, when we consider the policy mechanisms of interest in the area of technology management it is clear that the 'delivery' of policy is one of the most acute difficulties facing public authorities. This is especially clear in the case of firms which are least proficient in technology, where there is a lack of expressed demand for technology-related services. Firms are found to learn best from other firms with which they have established links, and by extension from agencies which adopt a market-orientated philosophy. It is therefore of great importance to use these relationships, whether they are geographically localized or not, to bring policies and programmes to bear on firms.

A better balance is needed between different types of competence targeted by policy. By far the largest proportion of public funds devoted to technology policy, in the European Union and elsewhere, falls into two categories: expenditure on the provision of what we have called the research and technology infrastructure, and expenditure on research subsidies. There are, of course, persuasive economic justifications for this, but it is certainly at odds with what this study suggests are the most urgent needs with regard to technology management on the part of the firms. It could be argued that these types of support—and research support for firms in particular—are relied on partly because of a lack of politically and operationally credible ways of supporting other (non-technical) competences directly. Because of this an increasing weight of indirect policy messages is packaged into collaborative research programmes, for example strong messages about how, why, and with whom to do research. These are important considerations which bear on but by no means exhaust the full scope of technology management in the firm.

Policy should work within and help to overcome the cognitive and organizational limitations of firms, which are reflected in bounded vision, constraints on absorptive capacity, and lack of expressed demand for innovation and technological change. At an immediate operational level, there is a need to find ways of accommodating these limitations within existing policy actions, and especially those addressed towards SMEs. No easy solution presents itself, but it seems clear that efforts should be focused especially on the development of measures to stimulate the demand for technology and technology-based services. Various approaches have been the subject of experiments at regional and national level, often supported through EU structural development, enterprise, and technology dissemination funds. They include: the encouragement of groups of firms and regional actors to develop joint outlooks on technology—e.g. Regional Innovation Strategies[2]—with the expectation that this will develop the confidence and contacts amongst the firms associated that are needed to solicit appropriate technology inputs; subsidies for technology audits by consultants who work for a short period with enterprises and help them clarify their technological competences and needs, as well as sources of possible supply; and 'teaching company' schemes, in which graduates are placed in firms for a period to carry out supervised projects and implant a greater awareness of technology opportunities.

Alongside existing technology policy measures it seems of clear importance to develop complementary approaches to address other

[2] A co-ordinated action supported by the European Union's Innovation programme and structural funds.

elements of technology-management competence. The key requirement is to extend existing policy frameworks and develop new ones directed towards improving the organizational capabilities of firms—individually and within networks—alongside developments in their technological competences. The European Union provides a favourable context for experiments in this area because of the broad reach and international nature of the EU's own research, technology development, and innovation programmes and the diversity of national-level programmes and related activities. A considerable amount of experimentation is indeed already taking place, and the most urgent need is to evaluate on a comparative basis the lessons which have been learnt, disseminate them widely, and develop from them more concentrated and targeted actions. It must of course be recognized that the organizational and management aspects of technological change are influenced by a wide variety of economic, institutional, and cultural factors, which themselves are addressed by policies far wider in scope than those associated with RTD and innovation. But within the specific context of technology policy, new approaches need to be developed to address organizational issues that arise in technology management at three levels.

To assist firms in developing and applying new models of organization and management which are of particular relevance to technology management

The solution to this is not just to divert funds from scientific and technological research programmes towards organizational research. Problems of uptake—difficult enough in conventional research programmes—loom large because of the specific character of organization and management and the private context of organizational learning. What policy needs to address is how to stimulate a better circulation of organizational knowledge between the public ('generic') and private spheres. The means to do this will differ from those appropriate to science and technology because of the different nature of the knowledge involved. Generic organizational and management knowledge exists in the form of abstract frameworks and techniques for understanding, structuring, and controlling organizations, or models of work organization, such as 'self-regulating work teams' or 'business process reengineering'. It emerges not from laboratories but from experience of organizational practices and innovations within working organizations, subsequently identifying a wider application of certain features of what has emerged and finding ways to customize it to new circumstances (Holti and Neumann, 1996). Management consultants and other business services are typically the agents by which, in the first instance, organizational knowledge is collated, generalized, packaged, and disseminated. However, academic institutions are the main source of systematic knowledge in organization and management relevant to technology, through both research and training (e.g. M.B.A. and Ph.D. programmes).

One important objective of technology policy in this area must be to improve the interactions between business services and industrial firms in the area of organizational and management aspects of technology-management competence. However policy needs to go beyond activities concerned with identification and transfer of best practice. Technological and organizational innovations take place in interaction with one another, even sometimes in the earliest stages of technology development. There is growing awareness in Europe of the importance of wider use by enterprises of the stock of academic knowledge and capabilities in the area, and of efficient application by industry of specific research results. At the same time, universities and business schools which are developing an interest in the area need to forge partnerships with firms, so as to generate data for their research and clients for associated services such as training. The tendency to separate management education and research from science and engineering, in the business schools as well as universities, could imply, however, that the academic sector is institutionally ill-equipped to deal with the very crucial current concerns which straddle the two disciplines.[3] Moreover, even when the conditions within academia are right, there are significant barriers in the way of establishing co-operative ventures with firms,

[3] In the USA, by contrast, one outcome of the national 'crisis' of competitiveness in the mid-1980s, was an intensification of the already well-developed interactions between the business and academic communities in the area of technology strategy and management, and its implications for economic development and competitiveness. Several highly publicized studies—*Made in America* (Dertouzos, Lester, and Solow, 1989) and *The Machine that Changed the World* (Womack, Jones, and Roos, 1990) for example—are evidence of this, as are the numerous academic centres established to carry out research on industrial technology management (MIT's International Centre for the Management of Technology or the 'industrial performance centres' established in various US universities, etc.).

relating both to lack of knowledge about good practice, as well as start-up costs and risks.

There is a general interest, then, in improving collaborations between the academic and business sectors in relation to the organizational and management aspects of technology management, and to do this on a Europe-wide basis. There are several practical ways in which this can be done within the framework of existing technology policies, by further developing and generalizing approaches which have already been pioneered. First, European technology management training courses could be promoted through industry–university partnerships, as in the EU's Comett scheme.[4] Second, there is scope for expanding successful approaches to collaborative research and dissemination programmes which promote organizational and management innovation in the context of technological change. Several experiments have been conducted at member-state level[5] based on 'action research' approaches and at member-state and EU level using the concept of 'business processes'.[6] The objective should be to fill an evident gap in technology-orientated research by means of instruments implemented in parallel with existing research projects to produce frameworks and generic models of organization and work practice directly relating to the technologies in question and facilitating their development, configuration, and uptake (Holti and Neumann, 1996). Finally, as indicated in Chapter 3, local and regional collaborative ventures such as technology-management 'clubs', associating business and management schools with groups of firms sharing similar management concerns, could be promoted and linked together at the European level.

To address the complexity and interface aspects of technological change in a market-orientated context

The contemporary characteristics of technological advance discussed earlier—multidisciplinarity, science linkage, innovation through technology integration, etc.—accompanied by the rising

[4] Now incorporated into the 'Leonardo' programme.

[5] One can cite, for example, the German AuT (Work and Technology) Programme, the Swedish INPRO (Integrated Product Development) and the UK MOPS (Management, Operations, People and Systems) programmes.

[6] For example in the UK, the Innovative Manufacturing Initiative and in the EU's Information Technology Programme the research activities targeted towards 'technologies for business processes'.

specialization of firms, has created an explosion of interface management problems. A structural change in research infrastructures and in the preoccupations of research programmes is needed away from the linear logic of science–technology–innovation which is still embedded in most research and technology-transfer programmes and related public-sector organizations, so as to give more emphasis to the 'downstream' aspects of market formation, the backward linkages to knowledge creation, and the horizontal linkages associated with the integration of scientific and technological disciplines.

The increasing weight being given to collaborative research in government programmes—long-established, of course, at the European level—must be considered as very positive. However there is still progress to be made to accommodate fully the crucial interactions between market formation and technological development—as expressed, for example, in the emergence of product conventions and *de facto* standards which reflect the most successful configurations of technologies in complex but previously unrealized technological products and systems. This is at least partly due to the persistence of traditional notions of the public good (embodied, for example in the concept of 'pre-competitive research') which still govern public initiatives.

But strenthening the link between the development of complex technologies and market formation also raises difficult questions of a purely practical nature. For example, can the appropriate organizational linkages between firms, and with other market actors, be effectively stimulated other than through technology promotion programmes? Are there alternatives to 'strategic' programmes which make strong assumptions about the development of technologies, with attendant risk of error ('picking winners')? Can technological determinism be avoided by targeting generic functions (i.e. functions associated with a wide range of application domains) rather than technologies? Answering these questions will involve serious comparative evaluation of various experiments that thave taken place over recent years in the EU context[7] and elsewhere.

[7] For example, one could cite the Open Microprocessor Initiative and other 'focused clusters' within the Information Technology Programme, sectoral initiatives within the Telematics Applications Programme and, most recently the European Commission's Research Industry Task Forces.

To assist the development of strategic orientations by firms, and setting the agenda of technological change
This involves taking account of both scientific and technological opportunities as well as expected evolution of demand and social welfare considerations. The multiple interdependencies between technologies, product and market development, and industrial organization impose a strong need for co-ordination between firms and other actors at the level of business strategies and technology management which, on our analysis, go well beyond what the market alone is capable of delivering. Providing a basis for improved co-ordination at these levels should not have the effect of limiting competition, but should facilitate the individual efforts of firms to cope better with the uncertainties over the long-term technology outlook and to anticipate better specific market and technological opportunities in the context of regulatory constraints and ever-shortening product and process life-cycles. 'Visioning' and 'foresight' type processes are already emerging at various levels—local, regional, sectoral, national, and European—on a decentralized basis and involving firms and other relevant innovation actors (OST, 1995; Grupp, 1994; Cuhls and Kuwahara, 1994). A quite general observation is that it is the *process* (e.g. the building of partnerships amongst different constituencies), not the final report, of such activities which is of most value to firms, hence the main policy need is to provide a 'virtual infrastructure' to stimulate and accommodate these activities and foster interactions between them. Besides action to promote the systematic development of this type of activity involving as wide a constituency as possible, support needs to be provided to ensure appropriate flows of information about broader technology and market developments, and to avoid inward thinking. In evaluating and drawing lessons from these activities a central concern must be to understand the right balance to encourage a better-informed outlook and dissemination of knowledge without overloading hard-pressed firms with unnecessary demands for their attention.

The approach sketched above is obviously consistent with the generally observed movement towards a more integrated approach to technology policy, on which we have remarked earlier. It argues for the use of instruments which are adapted to the conjuncture between market, technology, and organization. It argues strongly for structures and instruments to accommodate the inter-enterprise dimension of technology management. It argues for a more adaptive approach from the public authorities, as a partner to private enterprise in the adjustment to changing markets and changing technologies.

Appendix A: Study Participants

Research Consortium

Ben Dankbaar
Herman de Rooij
Jan Cobbenhagen
Robert Hassink
Fabienne Corvers
G. Philips

MERIT—Maastricht Economic Research Institute for Innovation and Technology, University of Limburg, Maastricht

Lena Tsipouri
G. Stamboulis

Department of Economics, University of Athens

Stefan Kuhlmann
Guido Reger
Markus Schroll
Uwe Gundrum
Hans Kungl

FhG-ISI—Fraunhofer-Institut für Systemtechnik und Innovationsforschung Karlsruhe

Rod Coombs
Albert Richards
Margarida Fontes

CROMTEC—Centre for Research on Organisations, Management and Technical Change, Manchester School of Management, University of Manchester Institute of Science and Technology

Francisco Garcia Blanch

P&G Soluciones Estrategicas, Madrid

Franco Malerba
S. Torrisi

Bocconi University, Milan

Thierry Suquet
Smaïl Aït Hadj
Isabelle Géniaux
Christian le Bas

Jorg Parczen
Groupe ESC Lyon, Lyon Graduate School of Business

Jean-Marie Jacques
Raymond Hordies
J. M. Mullenders
J. B. Charles
Faculté des Sciences Economique et Sociale, Université de Namur

Keith Pavitt
Roy Rothwell
Mark Dodgson
SPRU—Science Policy Research Unit, University of Sussex, Brighton

Panel of Industrial Experts

J. Balazard, Directeur Centrale Technique, Aerospatiale, Toulouse
C. Coggill, Finance Director and Company Secretary, Generics Group plc, Cambridge
R. de Pecher, Associazione Nazionale Costruttore Edili (ANCE), Rome
C. Floyd, European Associate Director, Arthur D. Little Ltd., London
A. H. Heins, formerly Vice President and Director of Research, Henkel, The Netherlands
J. Schmitt-Tiedemann, formerly Director of Research, Philips, Germany
B. Wafer, Director, Circa Group Europe Ltd., Dublin

Workshop Discussants

Professor T. Cronberg, Technical University of Denmark
Y. Farge, Chairman of the European Commission's Industrial Research and Development Advisory Committee (IRDAC)
B. Kramer, BMFT, Germany
H. Logue, European Commission, DGXII
N. C. Nielsen, Danish Technical Institute
H. Pero, European Commission, DGXII
I. Saragossi, European Commission, DGXII
Dr V. Corado Simoes, ISEG, Portugal
Professor K. Smith, STEP Group, Norway
B. Wicksteed, Segal, Quince, Wicksteed, UK

Appendix B: Study Reports

The following reports were all produced in the framework of the Monitor-Sast Study entitled *Research and Technology Management in Enterprises: Issues for Community Policy* published by the Office for Official Publications of the European Communities.

Conceptual Framework and Technical Guidelines, by B. Dankbaaret et al., MERIT, October 1993. Catalogue No. CG-NA-15426-EN-C (135 pp.—ECU 15.00).

Case Study on the Software Sector, by F. Malerba and S. Torrisi, Bocconi University, October 1993. Catalogue No. CG-NA-15427-EN-C (71 pp.—ECU 8.50).

Case Study on the Mechanical Engineering Sector, by G. Reger and H. Kungl, FhG-ISI, October 1993. Catalogue No. CG-NA-15428-EN-C (100 pp.—ECU 11.50).

Case Study on the Food Industries Sector in Less Favoured Regions of the European Community, by L. Tsipouri, et al., University of Athens, October 1993. Catalogue No. CG-NA-15429-EN-C (51 pp.—ECU 7.00).

Case Study on the Chemicals Sector, by J. Cobbenhagen and G. Philips, MERIT, October 1993. Catalogue No. CG-NA-15430-EN-C (89 pp.—ECU 10.00).

Case Study on Wallonie–Limburg–Aachen (*Euroregion*), by B. Dankbaar et al., MERIT, October 1993. Catalogue No. CG-NA-15431-EN-C (89 pp.—ECU 10.00).

Case Study on the Rhône–Alpes Region, by T. Suquet et al., Groupe ESC Lyon, October 1993. Catalogue No. CG-NA-15432-EN-C (67 pp.—ECU 8.50).

Case Study on England, by R. Coombs and A. Richards, CROMTEC, October 1993. Catalogue No. CG-NA-15433-EN-C (49 pp.—ECU 7.00).

Case Study on Spain, by F. Garcia Blanch, Soluciones Estrategicas P&G, Madrid, October 1993. Catalogue No. CG-NA-15434-EN-C (167 pp.—ECU 16.50).

Case Study on Portugal, by R. Coombs and M. Fontes, CROMTEC, October 1993. Catalogue No. CG-NA-15435-EN-C (56 pp.—ECU 8.50).

Case Study on Greece, by L. Tsipouri, University of Athens, October 1993. Catalogue No. CG-NA-15436-EN-C (55 pp.—ECU 7.00).

Problems and Issues in Research and Technology Management, by a panel of industrial experts, October 1993. Catalogue No. CG-NA-15437-EN-C (117 pp.—ECU 13.50).

Overall Strategic Review, by B. Dankbaar et al., MERIT, October 1993. Catalogue No. CG-NA-15438-EN-C (94 pp.—ECU 11.50).

Appendix C: List of Abbreviations

ANVAR	Agence Nationale pour la Valorisation de la Recherche
BIC	Business Information Centre
CAD	Computer-Aided Design
CAM	Computer-Aided Manufacturing
CAP	Computer-Aided Planning
CIM	Computer-Integrated Manufacturing
CNRS	Centre Nationale de la Recherche Scientifique
CRAFT	Cooperative Research Programme of the European Union
EU	European Union
HEI	Higher Education Institute
LFR	Less-Favoured Region
LPC	Local Production Community
MBA	Master of Business Administration
MINT	Managing the Integration of New Technology (an action under the EU's SPRINT Programme)
NC	Numerically-Controlled
OECD	Organisation for Economic Co-operation and Development
OST	Office of Science and Technology
PRC	Public Research Centre
RA	Research Association
RTD	Research and Technology Development
RTI	Research and Technology Infrastructure
SME	Small and Medium-sized Enterprise
SPRINT	Innovation and Technology Transfer Programme of the European Union
TP	Technology Park
TSF	Technology Strategy Forum
VALUE	Technology Diffusion Programme of the European Union

References

ABERNATHY, W. J., and UTTERBACK, J. M. (1978), 'Patterns of Industrial Innovation', *Technology Review*, 80/7: 40–7.

ACS, Z. S., and AUDRETSCH, D. B. (1990) (eds.), *The Economics of Small Firms: A European Challenge* (Dordrecht: Kluwer Academic Publishers).

ANDERSEN, E. (1991), 'HRD's Role in Concurrent Engineering', *Training and Development*, 47/6.

ARCHIBUGI, D., and PIANTI, M. (1992), *The Technological Specialisation of Advanced Countries* (Dordrecht: Kluwer Academic Publishers).

ATKINS (Atkins Management Consultants) (1990), *Strategic Study on EC Machine Tool Sector* (Epsom, Surrey).

BECKER, C. (1995), 'Evaluation of the Promotion of Information Technology: Microelectronics; Microperiphery and Microsystem Technology', in G. Becher and S. Kuhlmann (eds.), *Evaluation of Technology Policy Programmes in Germany* (Dordrecht: Kluwer Academic Publishers), 197–213.

BESSANT, J. (1993), 'The Lessons of Failure: Learning to Manage New Manufacturing Technology', *International Journal of Technology Management*, 8/2, 3, 4, special issue on *Manufacturing Technology: Diffusion, Implementation and Management*, 197–215.

BREINER, S., CUHLS, K., and GRUPP, H. (1994), 'Technology Foresight using a Delphi Approach: A Japanese–German Co-operation', *R&D Management*, 24/2: pp. 141–53.

BRÖDNER, P. (1993), *The Future of Industry in Europe* (*FINE*), discussion paper, IAT (Institut für Arbeit und Technik), Gelsenkirchen.

BURNS, T., and STALKER, G. (1961), *The Management of Innovation* (London: Tavistock).

CEST (Centre for the Exploitation of Science and Technology) (1995), *Bridging the Innovation Gap* (London: CEST).

CIRCA Group (1993), *Thematic Evaluation of the Impacts of CSFs for Research and Technology in Greece, Ireland and Portugal*, Synthesis Report (mimeo).

CLARK, K. B. (1985), 'The Interaction of Design Hierarchies and Market Concepts in Technological Evolution', *Research Policy*, 14: 235–51.

—— and FUJIMOTO, T. (1991), *Product Development Performance: Strategy, Organisation and Management in the World Auto Industry* (Boston: Harvard Business School Press).

COBBENHAGEN, J. and PHILIPS, G. (1993), *Research and Technology*

Management in Enterprises: Issues for Community Policy—Case Study on the Chemicals Sector (Luxembourg: Office for Official Publications of the European Communities), SAST Project No. 8; CG-NA-15430-EN-C.

COHEN, W. M., and LEVINTHAL, D. A. (1990), 'Absorptive Capacity: A New Perspective on Learning and Innovation', *Administrative Science Quarterly*, 35/1: 128–52.

COOKE, P., and MORGAN, K. (1990), 'Learning through Networking: Regional Innovation and the Lessons of Baden-Württemberg', Regional Industrial Research No. 5 (Cardiff: Regional Industrial Research).

COOMBS, R., and FONTES, M. (1993), *Research and Technology Management in Enterprises: Issues for Community Policy—Case Study on Portugal*, (Luxembourg: Office for Official Publications of the European Communities), SAST Project No. 8; CG-NA-15435-EN-C.

—— and HULL, R. (1995), 'BPR as IT-enabled Organisational Change: An Assessment', *New Technology, Work, and Employment*, 10/2: 121–31.

—— and RICHARDS, A. (1993*a*), 'Strategic Control of Technology in Diversified Companies with Decentralised R&D', *Technology Analysis and Strategic Management*, 5/4: 385–96.

—— —— (1993*a*), *Research and Technology Management in Enterprises: Issues for Community Policy—Case Study on England* (Luxembourg: Office for Official Publications of the European Communities), SAST Project No. 8; CG-NA-15433-EN-C.

—— SAVIOTTI, P., and WALSH, V. (1987), *Economics and Technical Change* (London: Macmillan).

CRAWFORD, M. (1992), 'The Hidden Costs of Accelerated Product Development', *Journal of Product Innovation Management*, 913.

CUHLS, K., and KUWAHARA, T. (1994), *Outlook for Japanese and German Future Technology: Comparing Technology Forecast Surveys* (Heidelberg: Physica-Verlag).

CYERT, R., and MARCH, J. (1963), *A Behavioural Theory of the Firm* (Englewood Cliffs, NJ: Prentice-Hall).

DANKBAAR, B., KUHLMANN, S., and TSIPOURI, L. (1992), *Research and Technology Management in Enterprises: Issues for Community Policy—Conceptual Framework and Technical Guidelines* (Luxembourg: Office for Official Publications of the European Communities), SAST Project No. 8; CG-NA-15426-EN-C.

DAVID, P. (1985), 'Clio and the Economics of QWERTY', *American Economic Review*, 75/2: 332–7.

—— and ROSENBLOOM, J. (1990), 'Marshallian Externalities and the Dynamics of Industrial Localisation', *Journal of Urban Economics*, 28: 349–70.

DERTOUZOS, M., LESTER, R., and SOLOW, P. (1989), *Made in America: Regaining the Productive Edge* (Cambridge, Mass.: MIT Press).

DIW (Deutsches Institut für Wirtschaftforschung) (1989), *Faserverbundwerkstoffe in der Mittelständischen Industrie* (Berlin/Aachen: DIW).

DODGSON, M. (1990), 'Technology Strategy in Small and Medium-Sized Firms', in Z. S. Acs and D. B. Audretsch (eds.), *The Economics of Small Firms: A European Challenge* (Dordrecht: Kluwer Academic Publishers), 157–67.

—— (1991), *The Management of Technological Learning* (Berlin: de Gruyter).

—— (1993*a*), *Technological Collaboration in Industry* (London: Routledge).

—— (1993*b*), 'Organisational Learning: A Review of Some Literature', *Organisation Studies*, 14/3: 375–93.

—— and ROTHWELL, R. (1994), *The Handbook of Industrial Innovation* (London: Edward Elgar).

DOSI, G. (1988), 'Sources, Procedures, and Microeconomic Effects of Innovation', *Journal of Economic Literature*, 26: 1120–71.

DUSSAUGE, P., HARTAND, S., and RAMANANTSOA, B. (1992), *Strategic Technology Management* (London: Wiley).

EDQUIST, C. (1994), 'Technology Policy: The Interaction between Governments and Markets', in G. Aichholzer and G. Schienstock (eds.), *Technology Policy: Towards an Integration of Social and Ecological Concerns* (Berlin: de Gruyter).

EHRLENSPIEL, K. (1991), 'Auf dem Weg zur Integrierten Produktentwicklung', *VDI Zeitschrift*, 133/3: 16–21.

ENSR (European Network for SME Research) (1994), *The European Observatory for SMEs, Second Annual Report* (Zoetermeer: EIM Small Business Research and Consultancy).

ESCHENBACH, R., HAFKESBRINK, J., and LÜTZ, S. (1995), 'Evaluation of the Promotion of Pre-Competitive Research and Development Cooperation in the Case of Microsystems Technology (MST)', in G. Becher and S. Kuhlmann (eds.), *Evaluation of Technology Policy Programmes in Germany* (Dordrecht: Kluwer Academic Publishers), 165–96.

European Commission (1994), *The European Report on Science and Technology Indicators 1994* (Luxembourg: Office for Official Publications of the European Communities), EUR 15897, CG-NA-15-897-EN-C.

EVERSHEIM, W., BRACHTENDORF, T., JACOBS, S., and KEHL, G. (1988), 'Strategische Entscheidungen im Werkzeugbau', *VDI Zeitschrift*, 130/6: 57–62.

FELDER, J., HARHOFF, D., LICHT, G., NERLINGER, E., and STAHL, H. (1994), *Innovationsverhalten der Deutschen Wirtschaft. Ergebnisse der Innovationserhebung 1993* (Mannheim: ZEW), Dokumentation No. 94-01.

FELSENSTEIN, D., and BAR-EL, R. (1989), 'Measuring the Technological Intensity of the Industrial Sector: A Methodological and Emprical Approach', *Research Policy*, 18: 239–52.

FINNE, H., LEVIN, M., and NILSSEN, T. (1995), 'Trailing Research. A Model for Useful Programme Evaluation', *Evaluation*, 1/1: 11–31.

FISCHER, M. M. (1990), 'The Micro-electronics Revolution and its Impacts on Labour and Employment', in R. Capellin and P. Nijkamp (eds.), *The Spatial Context to Technological Development* (Aldershot: Avebury).

FREEMAN, C. (1974), *The Economics of Industrial Innovation* (Harmondsworth: Penguin).

—— (1982), *The Economics of Industrial Innovation*, 2nd edn. (London: Pinter).

FRITSCH, M. (1995), 'The Market, Market Failure, and the Evolution of Technology Promoting Programmes', in G. Becher and S. Kuhlmann (eds.), *Evaluation of Technology Policy Programmes in Germany* (Dordrecht: Kluwer Academic Publishers), 311–29.

GARCIA BLANCH, F. (1993), *Research and Technology Management in Enterprises: Issues for Community Policy—Case Study on Spain* (Luxembourg: Office for Official Publications of the European Communities), SAST Project No. 8; CG-NA-15434-EN-C.

GARVIN, D. A. (1993), 'Building a Learning Organisation', *Harvard Business Review*, July–August, 78–91.

GEHRKE, B., and GRUPP, H. (1994), *Innovationspotential und Hochtechnologie. Technologische Position Deutschlands im Internationalen Wettbewerb* (Heidelberg: Physica/Springer).

GIBBONS, M., LIMOGES, C., NOWOTTNY, H., SCHWARTZMAN, S., SCOTT, P., and TROW, M. (1994), *The New Production of Knowledge. The Dynamics of Science and Research in Contemporary Societies* (London: Sage).

GRUPP, H. (1992*b*) (ed.), *Technologie am Beginn des 21. Jahrhunderts* (Heidelberg: Physica/Springer).

—— (1994), 'Technology at the Beginning of the 21st Century', *Technology Analysis and Strategic Management*, 6/4: 379–409.

—— (1995), 'Science, High Technology and the Competitiveness of EU Countries', *Cambridge Journal of Economics*, 19: 209–23.

—— and SCHMOCH, U. (1992), 'Perceptions of Scientification of Innovation as Measured by Referencing between Patents and Papers: Dynamics in Science-Based Fields of Technology', in H. Grupp (ed.), *Dynamics of Science-Based Innovations* (Berlin: Springer), 73–128.

HAFKESBRINK, J., and HORST, H. (1995), 'Evaluation of Technology Transfer Initiatives in the Framework of Innovation Support Programmes. Instruments of Technology Transfer Evaluation—The Case of Microsystems Technology', in G. Becher and S. Kuhlmann (eds.), *Evaluation of Technology Policy Programmes in Germany* (Dordrecht: Kluwer Academic Publishers), 197–213.

HAJEN, L., OPPOLZER, A., and STEUER, M. (1990), *Technische Entwicklung und Betriebliche Weiterbildung. Eine Empirische Untersuchung in der Metallindustrie* (Frankfurt/New York: Campus).

HAMMER, M. (1990), 'Re-engineering Work: Don't Automate, Obliterate', *Harvard Business Review*, 90/4: 104–12.

HILBERT, J., and SPERLING, H. J. (1993), *Die kleine Fabrik: Beschäftigung, Technik und Arbeitsbeziehungen* (Munich: Rainer Hampp Verlag).

HMSO (1994), *Realising Our Potential*, White Paper on Science and Technology (London: HMSO).

HOLTI, R. and NEUMANN, J. E. (1996), *Promoting Organizational and Management Innovation: Choosing Policy Instruments Relevant to Industrial Research and Technology Development Programmes* (Brussels: European Commission-DG XII), EUR 16748.

HUGHES, K. (1988), 'The Interpretation and Measurement of R&D Intensity—A Note', *Research Policy*, 17: 301–7.

JOHANSSON, H. J., MCHUGH, P., PENDLEBURY, A. J., and WHEELER, W. A. (1993), *Business Process Reengineering: Breakpoint Strategies for Market Dominance* (Chichester: Wiley).

JUSTMAN, M., and TEUBAL, M. (1995), 'Technology Infrastructure Policy (TIP): Creating Capabilities and Building Markets', *Research Policy*, 24: 259–81.

KODAMA, F. (1995), *Emerging Patterns of Innovation: Sources of Japan's Technological Edge* (Cambridge, Mass.: Harvard University Press).

KOGUT, B. (1993), *Country Competitiveness—Technology and the Organizing of Work* (Oxford: Oxford University Press).

KUHLMANN, S., and HOLLAND, D. (1995), *Erfolgsfaktoren der Wirtschaftsnahen Forschung* (Heidelberg: Physica/Springer).

—— and KUNTZE, U. (1991), 'R&D Cooperation by Small and Medium Sized Companies', in D. F. Kocaoglu and K. Niwa (eds.), *Technology Management. The New International Language*; PICMET '91, Portland International Conference on Management of Engineering and Technology, 27–31 October 1991, Portland, Oreg., USA, 709–12.

LANDABASO, M. (1995), *The Promotion of Innovation in Regional Community Policy: Lessons and Proposals for a Regional Innovation Strategy*, International Workshop on Regional Science and Technology Policy Research, Japanese Institute of Science and Technology.

LAY, G. (1993), 'Government Support of Computer Integrated Manufacturing in Germany: First Results of an Impact Analysis', *Technovation*, 13/5: 283–97.

LEIBINGER, B. (1992), 'Technologietrends der 90er Jahre im Maschinenbau', in (Stuttgart Arbeitsgemeinschaft der Industrie und Handelskammern in Baden-Württemberg, Industrielle Forschung—Motor des technischen Fortschritts), 30–8.

LUNDVALL, B.-A. (1992) (ed.), *National Systems of Innovation: Towards a Theory of Innovation and Interactive Learning* (London: Pinter).

MALERBA, F., and TORRISI, S. (1993), *Research and Technology Management in Enterprises: Issues for Community Policy—Case Study on the Software Sector* (Luxembourg: Office for Official Publications of the European Communities), SAST Project No. 8; CG-NA-15427-EN-C.

MARCH, J. G., and SIMON, H. A. (1993), *Organizations*, 3rd edn. (Oxford: Basil Blackwell).

MARTIN, B. (1991), *Research Foresight and the Exploitation of the Science Base* (London: HMSO).

METCALFE, S. (1995), 'The Economic Foundations of Technology Policy: Equilibrium and Evolutionary Perspectives', in P. Stoneman (ed.), *Handbook of the Economics of Innovation and Technological Change* (Oxford: Blackwell).

MEYER, M. H., and UTTERBACK, J. M. (1993), 'The Product Family and the Dynamics of Core Capability', *Sloan Management Review*, 34/3.

MITTELSTRAß, J. (1994), 'Grundlagen und Anwendungen—Über das schwierige Verhältnis zwischen Forschung, Entwicklung und Politik', *Forschung Entwicklung und Politik Chemie–Ingenieur–Technik*, 66/3: 309–15.

NELSON, R. R. (1992), 'The Roles of Firms in Technical Advance: A Perspective from Evolutionary Theory', in G. Dosi, R. Giannetti and P. A. Toninelli (eds.), *Technology and Enterprise from a Historical Perspective* (Oxford: Clarendon Press).

—— (ed.) (1993), *National Innovation Systems: A Comparative Analysis* (Oxford: Oxford University Press).

—— and WINTER, S. G. (1982), *An Evolutionary Theory of Economic Change* (Cambridge, Mass.: Harvard University Press).

OAKEY, R., ROTHWELL, R., and COOPER, S. (1988), *The Management of Innovation in High Technology Small Firms* (London: Pinter).

OECD (Organisation for Economic Co-operation and Development) (1988*a*), *Application of New technologies in Mature Industries. Synthesis of the Case Studies* (Note by the Secretariat) (Paris: OECD), DSTI/IND/GWP/88.3.

—— (1988*b*), *New Technologies in the 1990s: A Socio-Economic Strategy; Conclusions and Recommendations of a Group of Experts on the Social Aspects of New Technologies* (Paris: OECD).

—— (1992*a*), *Industrial Policy in OECD Countries: Annual Review 1992* (Paris: OECD).

—— (1992*b*), *Technology and Economy: the Key Relationships* (Paris: OECD).

—— (1992*c*), *Structural Change and Industrial Performance: A Seven Country Growth Decomposition Study* (Paris: OECD).

—— (1993*a*), *Technology Fusion: A Path to Innovation* (Paris: OECD).

OECD (1993*b*), *Small and Medium-sized Enterprises: Technology and Competitiveness* (Paris: OECD).

—— (1994), *Facts, Analysis, Strategies: Unemployment in the OECD Areas, 1950–1995* (Paris: OECD).

OST (Office of Science and Technology) (1995), *Progress Through Partnership: Report from the Steering Group of the Technology Foresight Programme* (London: HMSO).

PATEL, P., and PAVITT, K. (1994), *Technological Competencies of the World's Largest Firms: Characteristics, Constraints and Scope for Managerial Choice*, STEEP discussion paper No. 13 (Brighton: Science Policy Research Unit).

PAVITT, K. (1984), 'Sectoral Patterns of Technological Change: Towards a Taxonomy and a Theory', *Research Policy*, 13: 343–73.

—— (1992), 'Some Foundations for a Theory of the Large Innovating Firm', in G. Dosi, R. Giannetti and P. A. Toninelli (eds.), *Technology and Enterprise from a Historical Perspective* (Oxford: Clarendon Press).

PIORE, M., and SABEL, C. (1984), *The Second Industrial Divide* (New York: Basic Books).

PORTER, M. E. (1990), *The Competitive Advantage of Nations* (London: Macmillan).

PRAHALAD, C. K., and HAMEL, G. (1990), 'The Core Competence of the Corporation', *Harvard Business Review*, May/June 1990, 79–91.

PYKE, F. (1994), *Small Firms, Technical Services and Interfirm Co-operation* (Geneva: International Institute for Labour Statistics).

QUINN, J. (1993), *The Intelligent Enterprise* (New York: Free Press).

REGER, G., and KUNGL, H. (1993), *Research and Technology Management in Enterprises: Issues for Community Policy—Case Study on the Mechanical Engineering Sector* (Luxembourg: Office for Official Publications of the European Communities), SAST Project No. 8; CG-NA-15428-EN-C.

—— and KUHLMANN, S. (1995), *European Technology Policy in Germany* (Heidelberg: Physica/Springer).

—— CUHLS, K., and NICK, D. (1994), *Review of Best Management Practices and Tools for R&D Activities* (Karlsruhe: Fraunhofer Institute for Systems and Innovation Research).

ROBERTS, E. B. (1988), 'Managing Invention and Innovation', *Research Technology Management*, 31/1: 11–29.

ROSENBERG, N. (1991), 'Critical Issues in Science Policy Research', *Science and Public Policy*, 18/6: 335–46.

ROTHWELL, R. (1992), 'Successful Industrial Innovation: Critical Factors for the 1990s', *R&D Management* 22/3: 221–39.

—— (1993), 'The Fifth Generation Innovation Process', in K. H.

Oppenländer and W. Popp (1993), *Privates und Staatliches Innovationsmanagement* (Munich: Institut für Wirtschaftsforschung).

—— and Dodgson, M. (1992), 'European Technology Policy Evolution: Convergence towards SMEs and Regional Technology Transfer', *Technovation*, 14/4: 223–38.

Rothwell, R. et al. (1974), 'SAPPHO updated—Project SAPPHO Phase 2', *Research Policy*, 3/3: 258–91.

Roussel, P., Saad, K., and Erickson, T. (1991), *Third Generation R&D* (Cambridge, Mass.: Harvard University Press).

Ruivo, B. (1994), ' "Phases" or "Paradigms" of Science Policy?', *Science and Public Policy*, 21/3: 157–64.

Schienstock, G. (1994), 'Technology Policy in the Process of Change: Changing Paradigms in Research and Technology Policy?', in G. Aichholzer and G. Schienstock (eds.), *Technology Policy: Towards an Integration of Social and Ecological Concerns* (Berlin: de Gruyter).

Schmoch, U., Grupp, H., and Laube, T. (1994*b*), *Ziele einer Technologiepolitik der Schweiz: Standortvoraussetzungen und Technologische Trends* (Karlsruhe: FhG-ISI).

—— Breiner, S., Cuhls, K., Hinze, S., and Münt, G. (1994*a*), *Interdisciplinary Cooperation of Research Teams in Science-Intensive Areas of Technology* (Karlsruhe: Fraunhofer Institute for Systems and Innovation Research).

Schmoch, U., Hinze, S., Jäckel, G., Kirsch, N., Meyer-Krahmer, F., and Münt, G. (1993*a*), *Constraints and Opportunities for the Dissemination and Exploitation of R&D Activities: The R&D Environment* (Karlsruhe: Fraunhofer Institute for Systems and Innovation Research).

Schmoch, U., Strauss, E., Grupp, H., and Reiss, T. (1993*b*), *Indicators of the Scientific Base of European Patents* (Luxembourg: Office for Official Publications of the European Communities), EUR 15330; CG-NA-15-330-EN-C.

Schumpeter, J. A. (1942), *Capitalism, Socialism and Democracy* (London: George Allen & Unwin).

—— (1980), *The Theory of Economic Development* (Oxford: Oxford University Press).

Semlinger, K. (1993), 'Economic Development and Industrial Policy in Baden-Würtenberg: Small Firms in a Benevolent Environment', *European Planning Studies*, 1/4: 435–63.

Servatius, H. G. (1985), *Methodik des Strategischen Technologie-Managements* (Berlin: Erich Schmidt Verlag).

Smith, K. (1994), *Interactions in Knowledge Systems: Foundations, Policy Implications and Empirical Methods*, STEP Group (Studies in Technology, Innovation and Economic Policy) Report 10.84 (Oslo: STEP).

SMITH, P., and REINERSTEIN, D. (1991), *Developing Products in Half the Time* (New York: van Nostrand Rheinhold).

SOETE, L., and ARUNDEL, A. (1993), *An Integrated Approach to European Innovation and Technology Diffusion Policy*, Commission of the European Communities, SPRINT (Luxembourg: Office for Official Publications of the European Communities); CD-NA-15090-EN-C.

SPROULL, L., and KEISLER, S. (1991), *Connections: New Ways of Working in the Networked Organisation* (Cambridge, Mass.: MIT Press).

SUQUET, T., et al. (1993), *Research and Technology Management in Enterprises: Issues for Community Policy—Case Study on the Rhône–Alpes Region*, (Luxembourg: Office for Official Publications of the European Communities) SAST No. 8; CG-NA-15432-EN-C.

TEECE, D. J. (1988), 'Technological Change and the Nature of the Firm', in G. Dosi, C. Freeman, R. Nelson, G. Silverberg and L. Soete (eds.), *Technical Change and Economic Theory* (London: Pinter Publishers).

TÖDTLING, F. (1990), *Räumliche Differenzierung Betrieblicher Innovationen. Erklärungsansätze und Empirische Befunde für Österreichische Regionen* (Brussels: European Commission—DG XVI).

TROMMSDORFF, V. (1990) (ed.), *Innovationsmanagement in Kleinen und Mittleren Unternehmen. Grundzüge und Fälle—Ein Arbeitsergebnis des Modellversuchs Innovationsmanagement* (Munich: Verlag Franz Vahlen).

TSIPOURI, L. (1993), *Research and Technology Management in Enterprises: Issues for Community Policy—Case Study on Greece*, (Luxembourg: Office for Official Publications of the European Communities) SAST Project No. 8; CG-NA-15432-EN-C.

VAN DE VEN, A. H. (1988), 'Central Problems in the Management of Innovation', in M. L. Tushman and W. L. Moore (eds.), *Readings in the Management of Innovation* (New York: Harper Business), 101–22.

VON HIPPEL, E. (1976), 'The Dominant Role of Users in the Scientific Instrument Innovation Process', *Research Policy*, 5: 221–39.

—— (1988), *The Sources of Innovation* (Oxford: Oxford University Press).

VOORDIJK, H. (1994), *Naar integrale logistiek in bedrijfsketens. Ontwikkelingen in de bouw* (Maastricht: Universitaire Pers Maastricht).

WOLFF, H., BECHER, G., DELPHO, H., KUHLMANN, S., KUNTZE, U., and STOCK, J. (1994), *FuE-Kooperation von kleinen und mittleren Unternehmen. Bewertung der Fördermaßnahmen des Bundesforschungsministeriums*, Series Technik, Wirtschaft und Politik, vol. 5 (Heidelberg: Physica/Springer).

WOMACK, J. P., and JONES, D. T. (1994), 'From Lean Production to Lean Enterprise', *Harvard Business Review*, March–April, 93–103.

WOMACK, J. P., JONES, D. T., and ROOS, D. (1990), *The Machine that Changed the World* (New York: Rawson Associates).

Index

www.ingramcontent.com/pod-product-compliance
Lightning Source LLC
LaVergne TN
LVHW101441200926
848483LV00001B/16
* 9 7 8 0 1 9 8 2 9 0 2 8 5 *